IET TELECOMMUNICATIONS SERIES 56

Key Enablers for User-Centric Advertising Across Next Generation Networks

Other volumes in this series:

Volume 9	**Phase noise in signal sources** W.P. Robins
Volume 12	**Spread spectrum in communications** R. Skaug and J.F. Hjelmstad
Volume 13	**Advanced signal processing** D.J. Creasey (Editor)
Volume 19	**Telecommunications traffic, tariffs and costs** R.E. Farr
Volume 20	**An introduction to satellite communications** D.I. Dalgleish
Volume 26	**Common-channel signalling** R.J. Manterfield
Volume 28	**Very small aperture terminals (VSATs)** J.L. Everett (Editor)
Volume 29	**ATM: the broadband telecommunications solution** L.G. Cuthbert and J.C. Sapanel
Volume 31	**Data communications and networks, 3rd edition** R.L. Brewster (Editor)
Volume 32	**Analogue optical fibre communications** B. Wilson, Z. Ghassemlooy and I.Z. Darwazeh (Editors)
Volume 33	**Modern personal radio systems** R.C.V. Macario (Editor)
Volume 35	**Principles of performance engineering for telecommunication and information systems** M. Ghanbari, C.J. Hughes, M.C. Sinclair and J.P. Eade
Volume 36	**Telecommunication networks,** 2nd edition J.E. Flood (Editor)
Volume 37	**Optical communication receiver design** S.B. Alexander
Volume 38	**Satellite communication systems, 3rd edition** B.G. Evans (Editor)
Volume 40	**Spread spectrum in mobile communication** O. Berg, T. Berg, J.F. Hjelmstad, S. Haavik and R. Skaug
Volume 41	**World telecommunications economics** J.J. Wheatley
Volume 43	**Telecommunications signalling** R.J. Manterfield
Volume 44	**Digital signal filtering, analysis and restoration** J. Jan
Volume 45	**Radio spectrum management, 2nd edition** D.J. Withers
Volume 46	**Intelligent networks: principles and applications** J.R. Anderson
Volume 47	**Local access network technologies** P. France
Volume 48	**Telecommunications quality of service management** A.P. Oodan (Editor)
Volume 49	**Standard codecs: image compression to advanced video coding** M. Ghanbari
Volume 50	**Telecommunications regulation** J. Buckley
Volume 51	**Security for mobility** C. Mitchell (Editor)
Volume 52	**Understanding telecommunications networks** A. Valdar
Volume 53	**Video compression systems: from first principles to concatenated codecs** A. Bock
Volume 54	**Standard Codecs: image compression to advanced video coding, 3rd edition** M. Ghanbari
Volume 905	**ISDN applications in education and training** R. Mason and P.D. Bacsich

Key Enablers for User-Centric Advertising Across Next Generation Networks

Jose Simoes

The Institution of Engineering and Technology

Published by The Institution of Engineering and Technology, London, United Kingdom

The Institution of Engineering and Technology is registered as a Charity in England & Wales (no. 211014) and Scotland (no. SC038698).

First published 2013

The Institution of Engineering and Technology
Michael Faraday House
Six Hills Way, Stevenage
Herts, SG1 2AY, United Kingdom

www.theiet.org

British Library Cataloguing in Publication Data
A catalogue record for this product is available from the British Library

ISBN 978-1-84919-618-5 (hardback)
ISBN 978-1-84919-619-2 (PDF)

Typeset in India by MPS Limited
Printed in the UK by CPI Group (UK) Ltd, Croydon

To my parents: the foundation of my life,
my idols, my heroes

Contents

Preface

Telecommunication and Internet services are constantly subject to change, seeking the customer's full satisfaction. Enriching these services with innovative approaches such as context-aware, social, mobile, adaptable and interactive mechanisms, enables users to experience a variety of personalized services seamlessly across different platforms and technologies. In this sense, advertising is no exception, especially if we consider that it will become the business enabler for next generation services. Nevertheless, currently there is no cross-domain solution capable of delivering real-time advertising across heterogeneous environments or domains, and at the same time, addresses users needs, desires and intentions. This happens because most of the products available today are only used within isolated environments or silos. Therefore, managing advertising campaigns across different verticals is still very complex.

However, leveraging on the advances provided by Next Generation Networks (NGNs), together with the design principles inherent to Service-Oriented Architectures (SOAs) and capabilities offered by Service Delivery Platforms (SDPs), this scenario is about to change. Based on key conceptual entities called enablers, this work aims to change the current scenario. Enablers are software components, which are designed to address specific needs or requirements, but are generic enough to be re-used outside a single domain (through well-defined interfaces).

More concretely, this book introduces three distinct but complementary enablers. The Human Enabler (HE) provides a real-time context brokerage system capable of securely managing different types of user-related data in a standardized way. The Reasoning Enabler (RE) is the result of a well-defined methodology that enables new knowledge to be reasoned, based on previously stored data, by aggregating, correlating and inferring new information about people and their contexts. Lastly, the Session Management Enabler (SME) is responsible for abstracting the communication layers. It provides a context-aware multimedia delivery system capable of personalizing and adapting multimedia content according to a set of user and system pre-defined context data or rules, respectively. Altogether, they form the Converged User-Centric Advertising System (CUCAS) and introduce new features that address the needs of both users and advertisers.

The contribution of this research is fourfold: first, it analyzes and compares existing technologies and solutions from Information Technology (IT), telecommunications and advertising worlds. Then, based on the problems identified, a set of requirements is defined, an architecture is proposed and realized into a set of complementary enablers (CUCAS). The third point relates to the validation and evaluation of the developed prototype, which provides a better understanding of the current challenges around next generation advertising systems. Finally, based on the technological innovations, it promotes alternative business models that can change the way advertising is experienced by the end user. In fact, this book is the basis for Mobitto (www.mobitto.com), a commercial exploitation resulting from a spin-off initiative.

Abbreviations

3G	Third Generation
3GPP	Third Generation Partnership Project
4G	Fourth Generation
AAA	Authentication, Authorization, and Accounting
Ad	Advertisement
AI	Advertiser Interface
AIM	AOL Instant Messenger
AN	Advertising Network
AOL	America Online
API	Application Programming Interface
AR	Application Router
AS	Application Server
Axis	Apache EXtensible Interaction System
B2BUA	Back-to-Back User Agent
BEP	Bearer end-point
BIMs	Bearer Intermediaries
BSS	Business Support System
CAF	Common Advertising Framework
CGI	Common Gateway Interface
CMS	Content Management System
CPL	Call Processing Language
CPS	Common Profile Storage
CQL	Context Query Language
CRBE	Context/Rule-Based Engine
CRM	Customer Relationship Management
CSCF	Call Session Control Function
CtPD	Content Processor and Delivery
CUCAF	Converged User-Centric Advertising Framework
CUCAS	Converged User-Centric Advertising System

CxB	Context Broker
CxC	Context Consumer
CxP	Context Provider
DBSCAN	Density-Based Spatial Clustering of Applications with Noise
DVB	Digital Video Broadcasting
EPC	Evolved Packet Core
ETSI	European Telecommunications Standards Institute
FEC	Forward Error Correction
FEP	Flow end-point
FI	Future Internet
FMC	Fixed Mobile Convergence
GHP	Generic Human Profile
HCB	Human Context Broker
HDR	Human Data Repository
HE	Human Enabler
HPR	Human Profile Repository
HSS	Home Subscriber System
HTTP	Hypertext Transfer Protocol
ICQ	'I Seek You'
IdP	Identity Provider
IET	Institution of Engineering and Technology
IETF	Internet Engineering Task Force
IMEI	International Mobile Equipment Identity
IMS	IP Multimedia Subsystem
IP	Internet Protocol
IPTV	IP Television
ISC	IMS Service Control
IT	Information Technology
ITU-T	International Telecommunication Union, Telecommunication Standardization Sector
JCP	Java Community Process
LAN	Local Area Network
LTE	Long-Term Evolution
MBMS	Multimedia Broadcast Multicast Service
MDF	Media Delivery Function
MMS	Multimedia Messaging Service

MoC	Mobility Controller
MPLS	Multiprotocol Label Switching
MSN	Microsoft Network
MTO	Multiparty Transport Overlay
NAT	Network Address Translation
NGN	Next Generation Network
NoSQL	Not only SQL
NTL	Network and Transport Layers
OLN	Overlay Leaf Node
OMA	Open Mobile Alliance
ON	Overlay Node
OS	Operating System
OSA	Open Service Access
OSE	OMA Service Environment
OSI	Open Systems Interconnection
OSN	Overlay Source Node
OSS	Operating Support System
P2P	Peer-to-Peer
PCP	Privacy Context Provider
PDF	Profile Description Framework
PEEM	Policy Evaluation and Enforcement
PEL	Policy Expression Language
PoC	Push-to-talk over Cellular
PSTN	Public Switched Telephone Network
QoC	Quality of Context
QoE	Quality of Experience
QoS	Quality of Service
RAT	Radio Access Technology
REST	Representational State Transfer
RE	Reasoning Enabler
RTCP	Real Time Control Protocol
RTP	Real Time Protocol
RTSP	Real Time Streaming Protocol
SCIM	Service Capability Interaction Manager
SCF	Service Capability Function
SCP	Service Capability Functions
SCP	Social Context Provider

SDK	Software Development Kit
SDP	Service Delivery Platform
SDP	Session Description Protocol
SIP	Session Initiation Protocol
SM	Signaling Module
SME	Session Management Enabler
SMS	Short Messaging Service
SN	Social Network
SNA	Social Network Analysis
SOA	Service-Oriented Architecture
SOAP	Simple Object Access Protocol
SOM	Self-Organizing Map
SQL	Structured Query Language
SSO	Single Sign-On
SWOT	Strength, Weakness, Opportunity and Threat
TISPAN	Telecoms and Internet converged Services and Protocols for Advanced Networks
TMF	TeleManagement Forum
UA	User Agent
UAC	User Agent Client
UAS	User Agent Server
UMTS	Universal Mobile Telecommunications System
UN	User Network
UPSF	User Profile Server Function
URI	Uniform Resource Identifier
URL	Uniform Resource Locator
VM	Virtual Machine
WAP	Wireless Application Protocol
WAR	Web application Archive
WiMAX	Worldwide Interoperability for Microwave Access
WLAN	Wireless LAN
WSBPEL	Web Services Business Process Execution Language
WSDL	Web Service Description Language
XML	eXtensible Markup Language
XMPP	Extensible Messaging and Presence Protocol

Chapter 1
Introduction

1.1 Motivation

Users belonging to the digital generation are increasingly speaking of personalized service experiences. They want services that are interesting and relevant to them to be at their fingertips. The fact that users want to be the centre of their service world means it is possible to know a lot about them. In fact, this happens because people's future-oriented acts and thoughts are exposed to personal and collective interpretation and analysis, in an unprecedented manner, through various digital environments and channels. Personal futures become approachable and accessible when individuals store and share their life patterns and future-oriented thinking (intentions, hopes, fears, wishes, goals) in various ways and forms in digital environments. In this sense, the centre of the future gravity becomes visible and detectable through our physical and non-physical multi-linear existence (Koponen, 2010). On the other hand, different media environments and interaction methods create social spaces where one can observe the activity of others, taking place both in physical and in non-physical worlds. Figure 1.1 illustrates some of the data sources and consequent information that can be extracted from the digital world. However, they are willing to give information about themselves as long as privacy concerns are efficiently addressed (Chappell, 2008).

Moreover, with the rapid advance in technology, it is becoming increasingly feasible for people to take advantage of the devices and services in the surrounding environment to remain 'connected' and continuously enjoy the activity they are engaged in, be it sports, entertainment or work. Such a ubiquitous computing environment will allow everyone permanent access to services anytime, anywhere and anyhow (Baker et al., 2009). Therefore, being aware and able to communicate context is a key part of human interaction, particularly for mobile users, who can directly benefit from more personalized, adaptable, interactive and consequently more useful services. Still, due to the heterogeneity of current networks, platforms, technologies and devices, such a utopic vision is not yet a reality.

In this sense, advertising, as one of the oldest services available, is no exception to these trends. Thus, when empowered correctly, technology can provide the necessary tools to feed advertising systems with such human-aware knowledge. This will help to understand users' needs, desires and intentions, therefore raising the chances of advertising acceptance and, as a result, improving its perceived

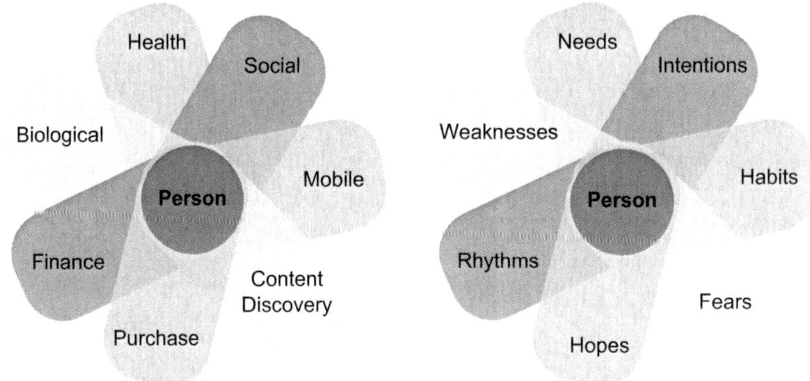

Figure 1.1 Personal sources of information vs. human behaviour prediction

Quality of Experience (QoE). In this book, QoE is defined as a consequence of a user's internal state (e.g. predispositions, expectations, needs, motivation, mood), the characteristics of the designed system (e.g. complexity, purpose, usability, functionality, relevance) and the context (or the environment) within which the interaction occurs (e.g. organizational/social setting, meaningfulness of the advertisement, voluntariness of use) (Hassenzahl & Tractinsky, 2006).

Nevertheless, despite the importance of the aforementioned principles, the realization of an advertising solution capable of addressing cross-domain targeting is still missing. This happens because most of the products available today are only used within isolated environments (e.g. television, radio, mobile, e-mail, etc.). Therefore, managing advertising campaigns across different verticals is still a very complex situation to accomplish. However, with the recent trend of Internet Protocol (IP)-fication among different domains, this scenario is about to change. During the last few years, it was possible to witness the deployment of Next Generation Networks (NGNs), aiming at enhancing telecommunication systems (Telco), from vertically separated application-specific networks, towards a unified horizontal layered IP-based network. Therefore, using IP as a common signalling protocol, it is possible to provide a homogeneous control layer, and consequently enable integrated Internet, telecommunications, media services and network convergence (e.g. Fixed Mobile Convergence (FMC)) (Blum et al., 2009).

Alongside this evolution, we can observe the adaptation of platform fundamentals from information technology in the service layer, where formerly closed systems are now adopting a flexible set of design principles and packaging functionality as a suite of interoperable services, which can be used within multiple separate systems from several business domains. This is commonly referred to as the adaptation of Service-Oriented Architectures (SOAs). Making the bridge between NGNs and SOA principles, Service Delivery Platforms (SDPs) are usually used to provide service control, creation, orchestration and execution environments, as well as abstractions for media control and other low-level communication capabilities.

Grounded on these premises, service enabler architectures are being adopted to provide abstracting toolboxes, allowing rapid deployment of advanced services within converged environments. This trend is mainly led by the Open Mobile Alliance (OMA), which specifies market-driven service enablers with the purpose of achieving interoperability across devices, geographies, service providers, operators and networks, while allowing businesses to compete through innovation and differentiation. These components can act either as stand-alone or in combination. The latter usually occurs within a service environment that leverages the interaction between different interoperable enablers. Some enabler examples specified by OMA include: Presence SIMPLE; XML Document Management (XDM); Policy Evaluation, Enforcement and Management (PEEM) architecture and Mobile Advertising (MobAd).

Having that said, when transposing these concepts to the current advertising industry, several of the existing challenges can be tackled. At the network and transport layers, it will be possible to see a convergence between Telco, Internet, media and entertainment environments, as these are the key industries for multimedia-based advertising (the focus of this book). This can be achieved by using a unified NGN environment at the architecture level. From the service-layer perspective, it is expected that multi-channel advertising can be managed in a seamless way. For this purpose, a Common Advertising Framework (CAF) (Figure 1.2) needs to be developed. To allow a better comprehension, Figure 1.2 depicts the role of a CAF, towards a future converged advertising architecture.

Based on the aforementioned technological innovations, the business side of advertising is also directly impacted. Having grown alongside with the expansion of the economy and being available through a myriad of channels, today, advertising is one of the most profitable industries (Hudson, 2008). In fact, according to GroupM (2010), global advertising spending in measured media was expected to exceed $500 billion for the first time ever in 2011. A more accurate overview over the advertising spending and related facts can be found in Appendix A. Finally, although technology does not change the roles of the different entities within the value of chain, it may open new opportunities to some of them. Operators might become prime publishers, while application developers can now enter in a market previously dominated by other bodies and consequently, disrupt existing revenue and business models for the advertising world.

1.2 Problem formulation

Several questions arise in complex advertising scenarios. However, this work focus has a well-defined goal.

> How to build a generic platform, capable of delivering real-time advertising across heterogeneous environments or domains, and at the same time, address users needs, desires and intentions?

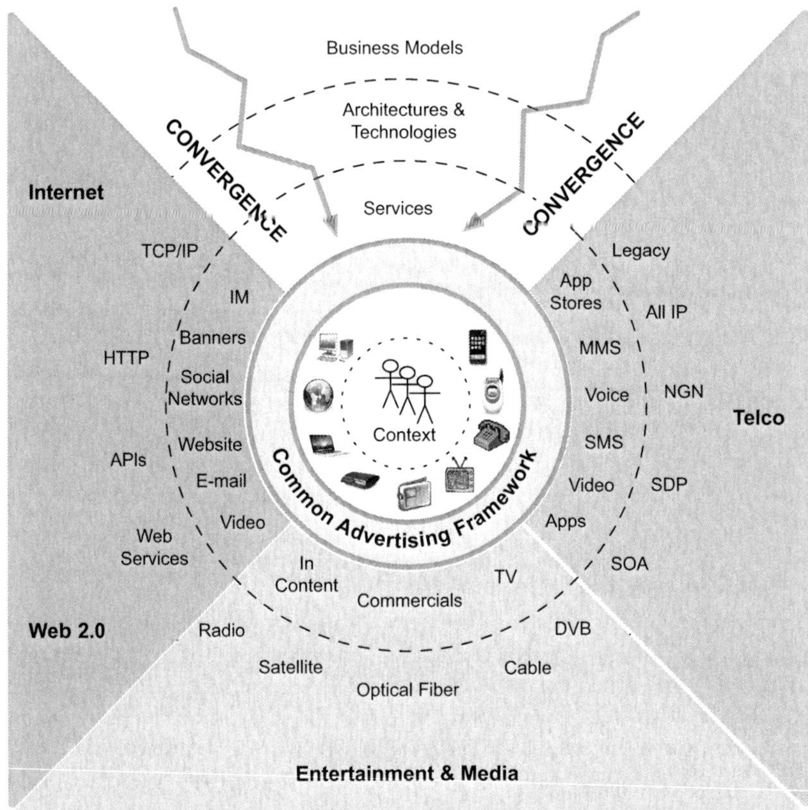

Figure 1.2 Overview of current advertising industry and the role of a common advertising framework

To allow a better understanding, the aforementioned research statement can be further divided into the following research problems:

Problem 1.1: Identify and understand what are the user concerns, needs, requirements or expectations for advertising in the future.
When designing architecture, it is a common practice to build up a list of requirements in the first place. Therefore, if this work aims to improve the way advertising exists today, it is necessary to understand what users are expecting, what their concerns are and what the most desirable characteristics are that this ecosystem should possess.

Problem 1.2: How to manage the required information (user, context, system) to efficiently address advertising?
Due to the amount and heterogeneity of systems and technologies existing nowadays, it is very difficult to access user-related data in a standard way. Furthermore, even when accessible, this information is usually dispersed over a myriad of

entities. Consequently, it is necessary to find a system capable of harmonizing and managing these procedures while assuring security, privacy and trust, allowing users to stay in control of their own data, disclosing it based on contextual policies.

Problem 1.3: Is it possible to understand and predict future human behaviours, desires, needs or intentions?
Every industry field that deals with people needs to know more about them in order to improve its offers. Advertising is no exception. Due to the amount of information available about people, shouldn't it be possible to model human behaviour, even if in a very small scale (limited to a specific actuation area)? And to predict it? Enabling this concept would allow changing the way advertising is perceived today, transforming it into a powerful recommendation service.

Problem 1.4: What is a suitable system architecture for converged advertising built on generic multi-access platforms?
Flexible management of multiple access technologies comprises the functionality to reallocate user sessions between different access systems. It is desirable that such an access handover provides seamless service continuity to the user. Furthermore, it is important that multimedia delivery is done in a personalized and adaptable way, taking into account innumerous sources of context-aware information. This solution should be able to optimize and balance both user satisfaction and network efficiency.

Problem 1.5: In which way can technological advances impact existing business and revenue models around advertising?
Most of the times, not only technology interferes with user experience, but it is usually limited by the way services are presented to a community. For this reason, by analysing the latest technology developments (especially the ones to be developed within this work), it is important to understand if the existing revenue and business models related to advertising can evolve and proportionate a new level of user experience.

1.3 Scope

While advertising can be experienced in a myriad of forms, this book focuses on a very specific subset: *online*, *real-time* and *dynamic*. This means that the channels being addressed must support some sort of connectivity to an NGN. Although the term NGN can be associated with different meanings (see section 2.2) here, it represents a multi-service IP network, providing an abstraction layer to any kind of networked service. Therefore, depending on the technology used at the control layer, it may even support service integration with circuit switched networks. Moreover, the content should not be static, that is, should be adaptable according to a set of contexts. Finally, it needs to be compliant with the constraints imposed by real-time requirements. To better understand this concept, Figure 1.3 highlights the scope of this work (a more detailed taxonomy is provided in Figure 2.4).

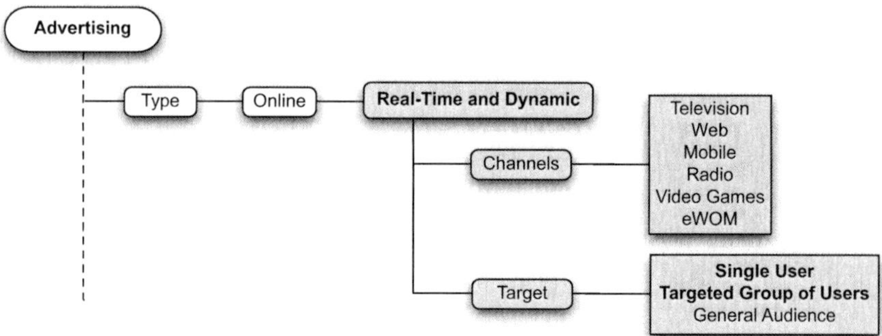

Figure 1.3 Scope within the advertising industry

The main reason for this choice is explained by the fact that looking at the present, but thinking about the future, real-time advertising is the one offering the best perspectives. In fact, independent of the channel being used, due to its nature, it allows advertisements to be adapted and contextualized to a specific moment in time and space. The ones identified here are television, radio, web, mobile, video games and electronic word of mouth. Nowadays, all of them allow mass market advertising, where the message/content reaches a general audience. Nevertheless, with the advance of technologies, this general audience can be segmented into smaller targeted groups as all of the aforementioned channels offer some sort of content, which itself is already targeted (contextualized) to a specific subset of people (e.g. cartoons are usually seen by children, soap operas by women and fighting sports by men).

Another key aspect inherent to this book is the *focus on the users* and not the advertisers themselves. This happens because most advertising solutions tend to address advertisers' needs (e.g. metrics) and rarely concern with the users' experience. This work, on the other hand, inverts these priorities, and aims at designing a solution that provides advertisements that users want, need and agree upon. For a better understanding of the aforementioned ideas, the following sections detail how this can be achieved, as well as, how this work contributes towards this vision.

1.4 Methodology

In order to tackle the problems related with advertising, it is important to understand to whom it is being targeted, as well as what are the current approaches, technologies and trends being applied. Based on this, the best architectural approach needs to be designed and consequently implemented. Then, through a set of iterative and repetitive validation methods, the same is improved and evaluated. Finally, the advertising system, or parts of it, is disseminated in a myriad of forms. For a better comprehension of the methodology adopted, Figure 1.4 highlights the aforementioned steps, which are listed below in more detail.

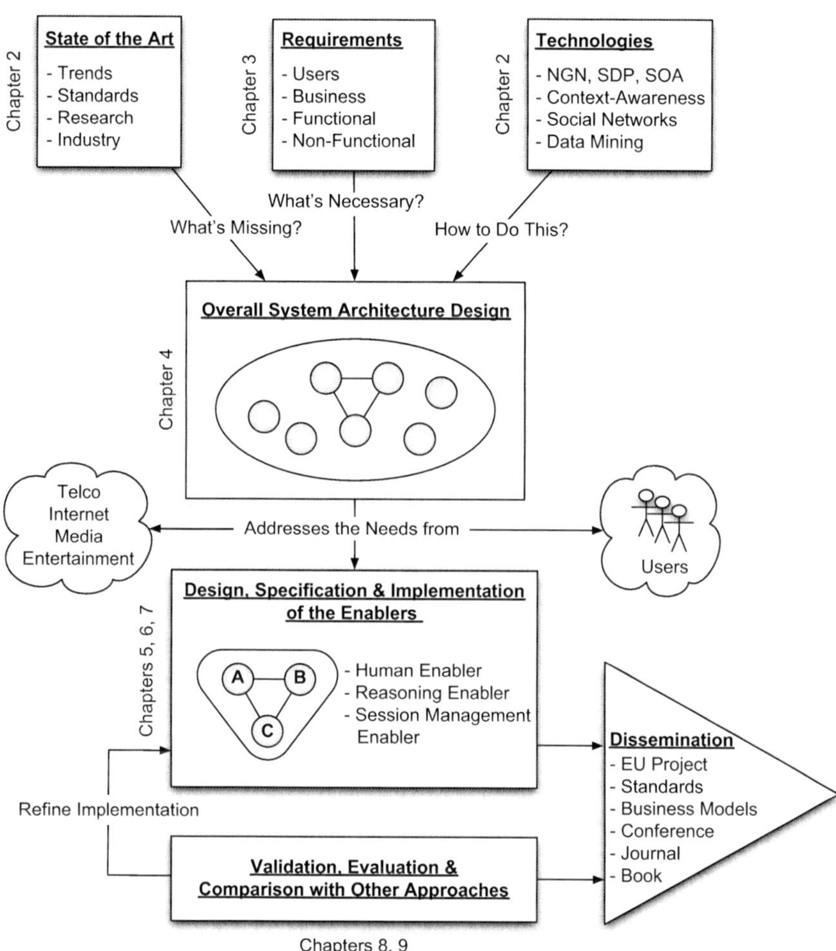

Figure 1.4 Methodology adopted to solve the identified advertising problems

1. Study and analyse existing service and architectural trends in the advertising domain. These include Telco, Internet, media or entertainment environments. Then, classify them into standards, products and research work.
2. Based on missing contributions, together with known advertising requirements (Eriksson & Akesson, 2008; Ranganathan & Campbell, 2002), it is important to understand what will be the key drivers for a better advertising framework. These are then defined as the system requirements.
3. Identify the set of technologies that can harness the design of the advertising framework. These will include contributions coming from different layers and domains.

4. Model an end-to-end advertising framework capable of incorporating the aforementioned technologies. Obviously, from all the concepts and enablers identified, only a few will be used in this approach.
5. Define, specify and implement the components identified. This includes the interfaces between the different entities, as well as the ones towards third-party providers.
6. *Validation of the proposed advertising framework*. The first step assures that the initial user, business and functional requirements are met. On a second stage, the architecture is evaluated through a set of iterative steps, until reasonable values for the non-functional requirements are achieved. Finally, this work is compared with other existing approaches.
7. *Work dissemination*. Although some of the components developed will only be tested here, others will be integrated in the scope of other projects. Likewise, several papers are to be published regarding the different phases of the research work.

In other words, to tackle the problem (and sub-problems) earlier identified and putting it in the context of the previously highlighted scope, this work separates the three main technological issues and addresses them independently inside dedicated enablers. In Figure 1.4, it is possible to see the relationship between the problems identified and the enablers used to handle these same issues. In here, an enabler refers to a software component that is designed to address a specific set of requirements, needs or operations. Furthermore, through a set of well-defined interfaces, the internal services can be used as stand-alone or exposed and accessed by third-party providers/applications.

There are many reasons for this approach, but the most preponderant one is its flexibility. On the technical side, it allows addressing the specifics of each problem in a more precise way. Moreover, this decision impacts positively most of the non-functional requirements (see Chapter 3). However, the advantages do not stop here. By adopting such strategy, the amount of scenarios the developed solution supports is much higher, as the enablers (components) can be reused outside the scope of advertising. This is also truth for the business models as different players can now enter in markets previously not accessible through existent solutions.

Altogether, three different enablers will be developed, creating an advertising toolbox capable of addressing the needs of a converged user-centric advertising platform that works within different types of NGNs. The following section provides more accurate information about the proposed solution and respective contribution towards the advertising industry.

1.5 Research contribution

With the purpose of improving user-perceived QoE for next generation advertising, it is important to understand what drives positive flows. Although the state-of-the-art suggests that adopting a user-centric strategy, combined with some sort of multimedia technology, is the path to go, this work goes one step further and identifies key principles that users identify as fundamental. These are personalization, contextualization, adaptation, privacy, mobility and interactivity.

Using these factors as common determiner, this research work explores a junction on the frontiers of three main domains: telecommunications, Internet and media/entertainment. Grounded on these principles, this book proposes – CUCAF – a Converged User-Centric Advertising Framework (a subset of a CAF introduced earlier), capable of addressing multimedia-based advertising, across heterogeneous environments. It is presented as a framework and implemented into three distinct but complementary components:

- *The Human Enabler (HE)* – a component that aims at improving the human side of information technology systems. It implements a context brokerage system capable of securely managing different types of user-related data (e.g. preferences, devices, contexts) in a standardized way. Furthermore, it allows this information to be requested or subscribed, under user self-defined context-aware privacy policies, in real-time.
- *The Reasoning Enabler (RE)* – the result of a well-defined methodology that enables new knowledge to be reasoned, based on previously stored data, by aggregating, correlating and inferring new information about people and their contexts. It is defined in a generic way and uses the aforementioned entity to expose the inferred data.
- *The Session Management Enabler (SME)* – a context-aware multimedia delivery system, capable of personalizing and adapting multimedia content according to a set of user and system pre-defined (or automatic) context data or rules. Among its tasks and together with a proper NGN infrastructure, it is responsible for abstracting the communication layers.

Altogether, when duly integrated, they will be able to tackle the problems identified in section 1.2. In fact, there is a direct correlation between these. The HE addresses user data management (Problem 1.2). The RE tries to improve human behaviour understanding (Problem 1.3), while the SME focuses on the multimedia delivery across heterogeneous networks (Problem 1.4).

As a consequence, new business models can arise (Problem 1.5). Nevertheless, this is only possible due to the requirements identified initially (Problem 1.1). Figure 1.5 depicts these relationships.

In parallel, some other components, extensions and modules were developed to address the missing interactions provided by the CUCAF. In this sense, the advertiser interface, social and privacy context providers are particularly relevant, as these provide the tools to interact with the advertising system. Although these have not been defined as enablers, they are implemented using the same philosophy, where its functionalities can be accessed by external entities through well-defined interfaces. To better understand the impact of this book, Figure 1.6 shows this work's contributions from a different perspective.

However, this book aims to go beyond technical aspects. This work also contributes to improve existing business models based on the technical innovations previously mentioned. A practical example is the spin-off company, Mobitto

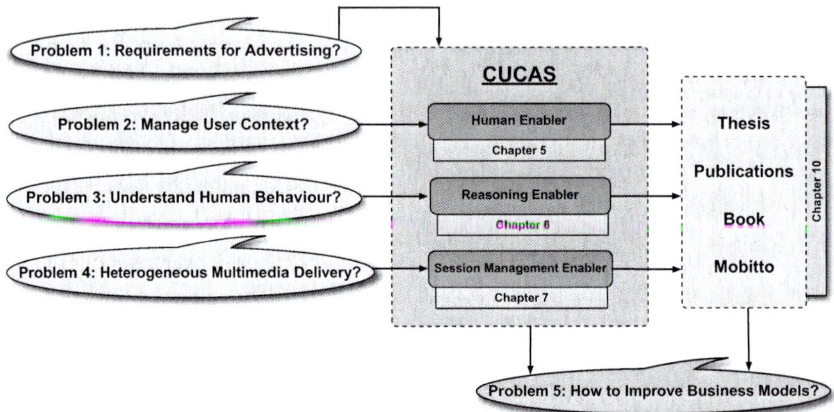

Figure 1.5 Summary of the problems vs. books contributions

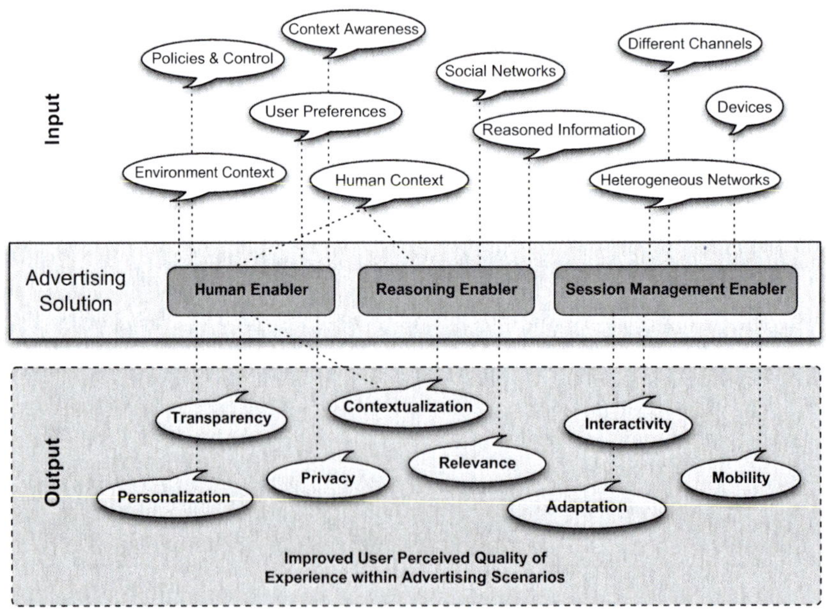

Figure 1.6 The input vs. output of the proposed advertising framework

(www.mobitto.com) that was created based on this book's foundations. Furthermore, it is important to mention that most of the technological developments hereby proposed are also applicable to other industries and not just advertising.

In terms of numbers, the work done on this book is a contribution of more than 25 scientific peer-reviewed papers, which are published as articles in journals, magazines and conference proceedings. A list of these is included with the references at the end of this book.

1.6 Outline

The remainder of the book follows a straightforward schema:

Chapter 1 – Introduction: explains the main motivation for this work, highlighting the role of advertising in next generation services. Then, it introduces the main problems to be addressed, and the adopted methodology to tackle them. In addition, it overviews the book and briefly outlines the content in the subsequent chapters.

Chapter 2 – State of the art: covers the basic aspects around NGNs, SOAs and advertising, namely their definitions and how they evolved over the times. Furthermore, it overviews related work at standardization, industry and research levels. Based on these, it discusses how this research relates to the specific fields.

Chapter 3 – Requirement analysis: With the purpose of solving the problems identified in Chapter 1 and focusing on the user-perceived QoE, it introduces the most important user requirements. Based on these, both the functional and non-functional requirements are identified to provide the next generation advertising ground functionalities.

Chapter 4 – Designing a converged user-centric advertising framework (CUCAF): details the experimental design and methodology, including a user-oriented survey, the system model and some functional entities. It also discusses the framework and enhanced advertising system architecture, focusing on its characterization, validation and evaluation.

Chapter 5 – Design and specification of the human enabler: describes the importance of having a standardized way of managing user-related data. In this sense, it presents the unified human profile, a flexible data structure capable of storing generic user-related context data. Moreover, it shows how this structure can be ported into an actual component capable of securely managing user-related context, the HE.

Chapter 6 – Design and specification of the reasoning enabler: highlights the importance of knowing more about the users in order to address their needs, desires and intentions. Furthermore, it presents the technical approach, the technologies used and the algorithms applied to allow both user behaviour modelling and inference. Lastly, it introduces the syntax to manage this information and presents some application scenarios.

Chapter 7 – Design and specification of the session management enabler: presents the architecture and system description responsible for managing and delivering multimedia advertising content across heterogeneous networks. It also mentions how this solution can achieve optimization and increase efficiency at the session, network and transport layers. Then, it exemplifies how, through a set of advertising use cases.

Chapter 8 – Implementation and testbed description: introduces and describes the components that realize the proposed Converged User-Centric Advertising System (CUCAS) framework, namely the human enabler, the reasoning engine and the session management enabler. In addition, it depicts the overall testbed scenario under which the components were further evaluated and validated.

Chapter 9 – Validation and evaluation: focuses on the evaluation and validation of each of the entities used in the realization of the advertising solution. This means that the user, functional and non-functional requirements defined in Chapter 3 are cross-checked against the functionalities of the deployed components. Furthermore, it analyses the possibility of offering new business and revenue models based on the work developed.

Chapter 10 – Conclusions and future work: concludes the book with a discussion of the current direction this technology may be taking society (and vice-versa). Moreover it theorizes potential ramifications of the proposed framework on a variety of academic disciplines and speculates on how the research will evolve in response to mentality changes of the general population.

Chapter 2
State of the art

Chapter 1 gave an overview of the motivation, problems being addressed, adopted methodology and main contributions of this book. This chapter takes advertising as main topic and covers its basic aspects, namely definition, how it has evolved over time, what types of advertising exist and how it can impact society, politics, culture and economy. Furthermore, it overviews related work at standardization, industry and research levels. Based on these, it discusses how the work in this book relates to the specific fields.

2.1 Basic concepts

2.1.1 What is advertising?

To some, advertising is an unnecessary assault on our senses, typified by a loud-mouthed boor, hard selling something we don't want or need. To others, it is an inconvenient economic burden, inflating the price of everything we buy. In a more generic definition, according to one commentator (Hudson, 2008), 'advertising is a form of communication intended to persuade its viewers, readers and listeners to take some action. It usually includes the name of a product or service and how that product or service could benefit the consumer, persuade potential customers to purchase or consume that particular brand'. In a similar definition, McChesney (2008) classifies advertising as 'a form of commercial mass communication designed to promote the sale of a product or service, or a message on behalf of an institution, organization, or candidate of a political office'.

But actually, advertising takes many forms, aside from the mass version, which defines the term today. It includes all human communication that overtly attempts to elicit or facilitate a transaction, action or reaction. Beginning where objective information leaves off and stretching until subjective art takes over, advertising is a hybrid of information and aesthetics that can be tracked back to the non-camouflage markings in animals (Blair, 1994). Due to its broadness, it has become such an important factor in the economy of most countries that according to McChesney (op cit), it also itself changes the economy, society, culture and the political system.

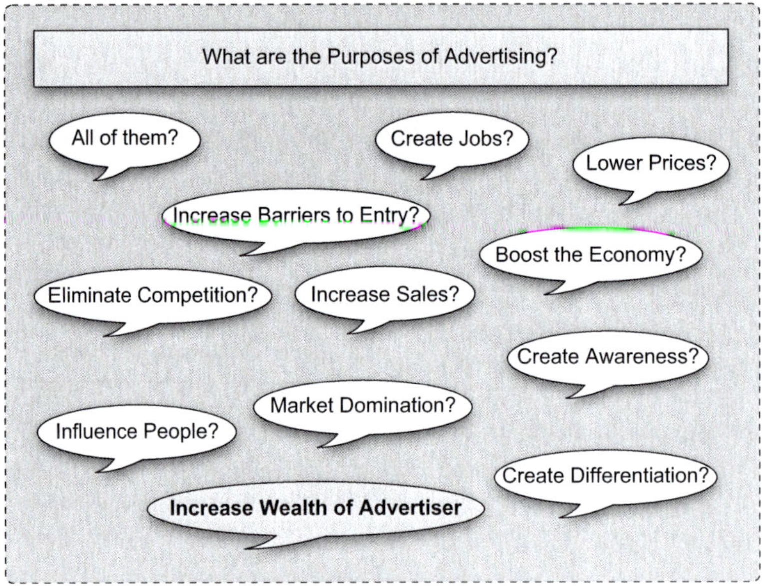

Figure 2.1 The purposes of advertising

Independently of the adopted definition, its purposes are very clear and similar in different areas. Figure 2.1 depicts a summary obtained from the previous literature.

For this work, we would like to define it as something different. Advertising will be seen as needed, useful and desirable. For the end-user, it will be perceived as a recommendation, or as defined in its infancy (see section 2.1.2), as a personalized word of mouth. It will respect people's privacy and may become a service itself, where users are offered with relevant content according to a personalized, adapted and contextualized set of interactions. It will become interactive, offering a myriad of immersive and compelling user-oriented experiences. Nevertheless, just like in the previously presented definitions, the main purpose will be increasing the wealth of the advertiser.

2.1.2 The evolution of advertising

Although some people disagree on how old advertising is, some authors (Blair, 1994; Tolani, 2008) believe it originated in 3000 bc, at least in a primitive form, as signs painted on a wall or rocks. Later discoveries mentioned that in ancient Rome and Pompeii, artefacts (some plates) were found announcing a property for rent and a tavern for travellers, respectively. In the medieval times, the early outdoor advertising competitors were town criers, employed by the royalty and merchants, to praise public announcements and goods (the word of mouth effect). However, some authors tend to believe that these pre-market societies did not use advertising, as during these times, humankind used to solve the two basic economic problems – how to allocate scarce resources and how to distribute output (wealth) – according to the principles of reciprocity (every member of a society knows his assigned tasks and receives his rightful share), redistribution (based on central authority who determines what to produce and what to get back) and house holding (produce for own consumption) (Norris, 1980).

Therefore, it was Gutenberg's invention of the moveable-type printing press in 1450 that boosted advertising, resulting in the mass production of posters and circulars. The first advertisement in English was a handbill printed in 1472. It was primarily an announcement of a prayer book for sale. Two hundred years later, the first newspaper Advertisement (Ad) appeared, offering a reward for finding 12 stolen horses. By the seventeenth century, classified ads were appearing frequently in England's newsweeklies. These ads featured simple descriptions of products and their prices. Nevertheless, as a side effect from the Industrial Revolution, consumption became the watchword. To encourage such consumption, the advertising industry has expanded, developing sophisticated techniques for inducing new desires and needs among ordinary people (Tarnas, 1993). Often using manipulation, sex appeal and other emotional inducements, advertisers have been able to attract people towards objects and services of which they never felt any need, before the new era of advertising emerged. Illustrations and colour appeared in advertisements in the late nineteenth century (Tolani, 2008). However, this sort of advertising did not provide any sort of interaction.

Later in 1920, with the appearance of radio, a new way of communicating finally arrived. Although limited to one way, the interaction could occur and advertisers could finally talk to their audience. In 1938, advertising in radios had already surpassed magazines and newspapers. Then, in 1940 came the first television advertisement, the Bulova watch, which initiated the era of visual broadcasting (radio allowed audio broadcasting). The market did not embrace television initially because of the high cost of TV sets and the lack of programming. However, as the economy improved in the 1950s, television's popularity surpassed that of radio. Soon, the industry considered it the No. 1 medium for advertising. Nevertheless the advertising revolution happened in the 1960s, where the industry became more scientific. This period witnessed some of the most creative ads of the time. Instead of focusing on the product, ads endeavoured to strengthen the brand and create an image of the company. Advertising also became subtler and more intelligent, often adopting a conversational style. In this sense, advertising turned into a major industry in the twentieth century. The ads at that time cleverly used all media, including newspapers, television, direct mail, radio, magazines, outdoor signs, and, of course, the Internet.

With the upcoming of the web, advertising could be differentiated from traditional media like television, radio and print in at least three important ways (McCandless, 1998):

- Every advertisement served on the web goes to an individual who can be characterized personally.
- A web ad receives direct feedback concerning its effectiveness (clicks).
- The real-time nature of the Internet lets ad campaigns adapt to nearly instantaneous feedback.

These methods have constantly evolved through the concept of contextual advertising, which in its infancy was simply defined as targeting advertisements through broad, pre-determined categories. Although this was certainly a fair attempt at attracting consumers, quite a few problems arose from this model. First, categories in general are broad and don't allow for granular targeting. Second, this model didn't allow for advertisements to perform on dynamic content, therefore greatly decreasing relevancy and performance for both the publisher and the advertiser. Forty per cent to fifty per cent of content on the web is dynamic, so an inability to target dynamic content inhibits the relevancy of the ads and, therefore, decreases the likelihood of sales. Third, under this model, optimization occurs at a manual level, which is not only time-consuming but also leaves room for error (Sears, 2005).

Giving the first steps in the world of advertising, context-awareness technologies started as a trial in early 2000s. However, in this period, the only context type being considered was location. These initial efforts have been reported in Yunos et al. (2003), together with the requirements, challenges and proposed business models. Despite the vision and pioneering efforts, none of the businesses resisted until today. The main explanation for this can be found in Chapter 3. Basically, at this time it was not possible to gather the necessary conditions to address both user and business parties' requirements.

This brings us to where we are today, where we can apply a combination of technologies to provide real-time, personalized, contextualized and adapted advertising. Assuming people want to be able to access their services (which will be merged with advertising) wherever they are, at any time and anyhow, it is very likely to see mobile advertising boosting in the upcoming years. In fact, this trend is confirmed by the numbers presented in Appendix A (Magma Global, 2011). Therefore, knowing more about the users and giving them control over their data will open unprecedented opportunities in this field. To sum up, Figure 2.2 depicts a summary of the evolution of advertising through different eras.

So, where is advertising taking us? As it was described, advertising has been experiencing incredible changes over the past decades and it is expected to continue in this direction. Aiming the full satisfaction of the end-user, it will definitely

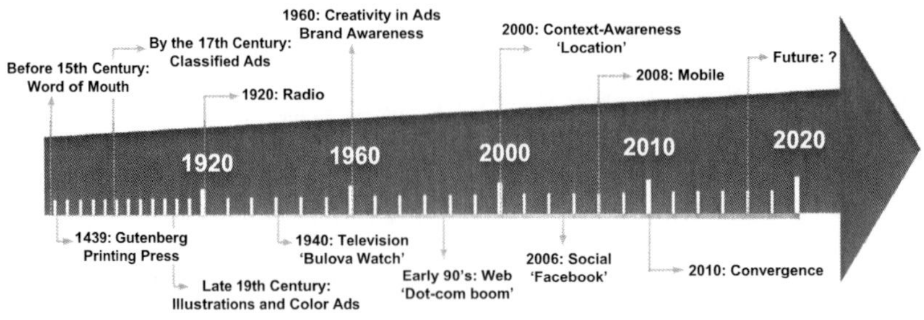

Figure 2.2 The evolution of advertising

need to adapt to the ubiquity of the world we are living in and enable a cross-device scenario and environment of advertising experience.

2.1.3 Types of advertising strategies

The concept of advertising was clearly explained in subsection 2.1.1. However, the fields of knowledge behind it were not mentioned. Furthermore, ads can take several forms and formats and are usually associated with different goals. This subsection will cover all these issues.

It is very difficult to narrow the number of subjects advertising is exposed to; nevertheless, Figure 2.3 depicts some of them, while the following text explains why. In few words, advertising can be seen as the result of multi-disciplinary research, which results into a science itself. To facilitate the comprehension, the different fields were divided into three categories. The first represents the *why*, and deals mainly with human-related sciences, namely, sociology, anthropology, psychology and history. These represent the way people think, react, live and perceive things. It is usually associated with the emotions. The second subset, the *what*, encompasses finance, marketing, mathematics and economy. Together, they define the goals, the strategy and the outcome of the campaigns. In addition, they deal with the rules and principles under which people are concerned. In other words, they provide the rational thinking. Lastly, the *who* basically represents the technological aspects allowing the elaboration and correct execution of the advertisement. Using key principles from both 'why' and 'what' sub-groups, this research is mostly visible on the technological side, realizing several issues identified in the entire multi-variable ecosystem that composes advertising, as it exists today.

Virtually any medium can be used for advertising. Commercial advertising media can include wall paintings, billboards, street furniture components, printed

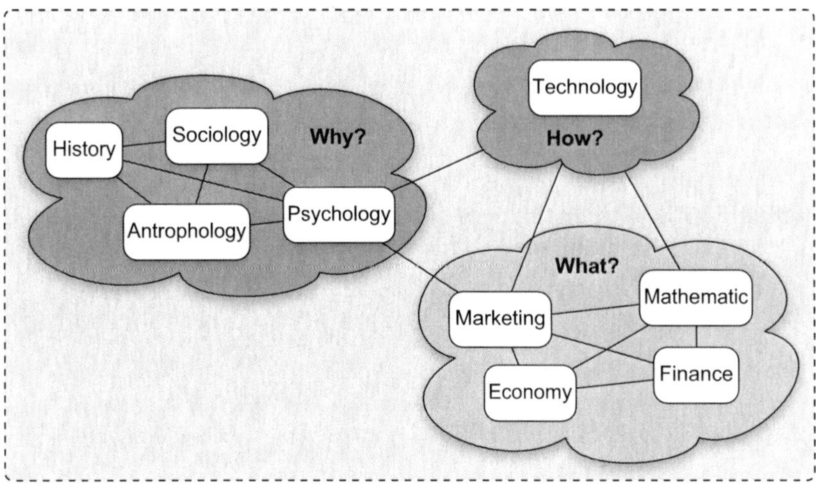

Figure 2.3 What are the influencing fields of Advertising strategies?

flyers and rack cards, radio, cinema and television adverts, web banners, mobile telephone screens, shopping carts, web popups, skywriting, bus stop benches, human billboards, magazines, newspapers, town criers, sides of buses, banners attached to or sides of airplanes ('logojets'), in-flight advertisements on seatback tray tables or overhead storage bins, taxicab doors, roof mounts and passenger screens, musical stage shows, subway platforms and trains, elastic bands on disposable diapers, doors of bathroom stalls, stickers on apples in supermarkets, shopping cart handles (grabertising), the opening section of streaming audio and video, posters and the backs of event tickets and supermarket receipts. In fact, any place an 'identified' sponsor pays to deliver their message through a medium is advertising.

Instead of focusing on the variety of existing medium of channels of advertising and show how these types appeared, in here, only a subset will be discussed. The best way to understand the methods or types of advertising is to distribute them into different categories. By presenting the most used channels existing nowadays, Figure 2.4 represents a simplified version of an advertising taxonomy. Furthermore, it identifies the scope of our work.

Looking at the present, but thinking about the future, real-time advertising is the one offering the best perspectives. In fact, independently of the channel being used, due to its nature, it allows advertisements to be adapted and contextualized to a specific moment in time and space. The ones we identified here were television, radio, web, mobile, video games and electronic word of mouth. Nowadays, all of them allow mass market advertising, where the message/content reaches a general audience. Nevertheless, with the advance of technologies, this general audience can be segmented into smaller targeted groups as all of the aforementioned channels offer some sort of content, which itself is already targeted (contextualized) to a specific subset of people (e.g. cartoons are usually seen by children, soap operas by women and fighting sports by men).

However, in the future, it is expected that a one-to-one marketing solution will be available. In this scenario, advertisements will be personalized to a single user. This does not mean that the same ad cannot be experienced by different users, but it will happen under different conditions. This vision can be labelled as the evolution of behavioural marketing, which is now enhanced with context. This is only possible due to technological advances, allowing personal, social and environmental context to be disclosed to advertisers.

Despite all these efforts, at the same time these advances occur, consumer electronic producers (manufacturers of the devices) and software engineers are developing ways to eliminate advertising from all the earlier mentioned advertising channels. A couple of examples are pop-up blockers and recorders with capability to remove ads (e.g. TiVo) among others. To fight back this trend, advertisements are now coming embedded into the content itself, typically inserted into black backdrops (in the case of television) or replacing scenes (i.e. audio, video or text), which are not relevant to the remote audience. Nevertheless, they are usually static. To improve the efficiency of these strategies, the same advertisement may be

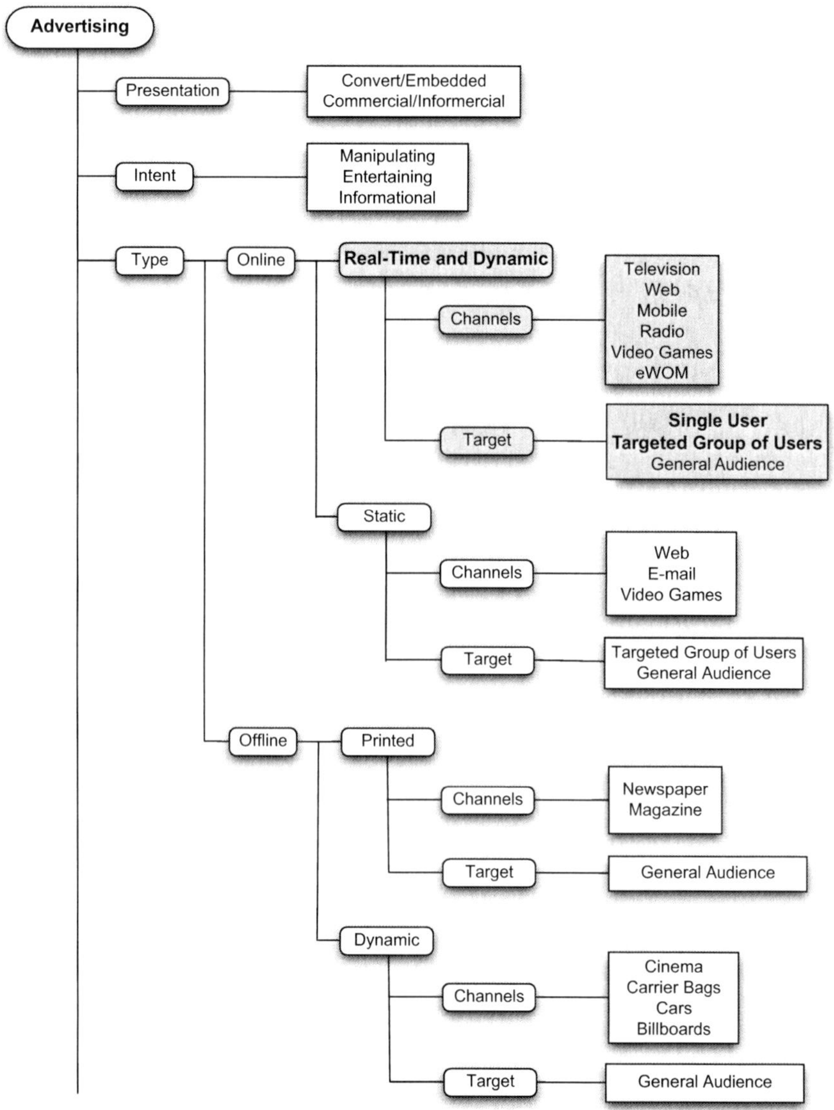

Figure 2.4 Taxonomy of advertising (scope of this work highlighted)

personalized and distributed in different forms and formats across different channels, engaging users in a more universal, but still personal experience.

But is this the best strategy? Are users satisfied with the ads they are receiving? Can they control what kind of ads they are receiving? Can they select who can advertise to them? Can they specify how? And when? The answer to all those questions is probably no. That is why it is important to re-think advertising strategies and introduce the Privacy, Permission and Preference (PPP) principle (Giselle, 2008), where users' interests come in the first place. Maybe it will increase advertisers' margins, maybe not, but it will definitely make this world a better place.

2.1.4 The impact of advertising

Despite the main goals of advertising have been defined in section 2.1.1, there are other side effects where advertising plays a preponderant role and has a great impact.

2.1.4.1 Economy

The stimulation for demand of products and services helps the economy to grow more stronger. New inventions become known much faster and can establish their spot in the sales figures of the economy. Actually, advertising has developed itself into a billion-euro industry on which many depend. In Germany for example, the advertising industry contributes with 1.5% of the gross national income and in 2010 employed a total of 160,000 people directly, plus 30,000 in related industries (Nicke, 2010/2011). Moreover, if there are more people buying products, the overall costs will drop and the product will become much cheaper for the customer, which raises his/her willingness to buy even more. On the other hand, advertisements are very expensive and some economists believe that these costs are put on top of the actual price paid by the customer. In fact, one of the ironies of advertising is that as commercialism increases, it makes it much more difficult for any particular advertiser to succeed, hence, entering a phase of cannibalism (McChesney, 2008).

2.1.4.2 Society

While advertising can be seen as necessary for the economic growth, it is not without social costs. Unsolicited commercial e-mails and other forms of spam are so prevalent that they become a major nuisance to the users of these services. Moreover, they are increasingly invading public spaces and the private sphere of people, resembling a new type of dictatorship that cannot be escaped. Unfortunately, the most important element of advertising is not information but suggestion, more or less making use of associations, emotions (appeal to emotion) and driving dormant in the sub-conscience of people, such as sex drive; herd instinct of desires, such as happiness, health, fitness, appearance, self-esteem, reputation, belonging, social status, identity, adventure, distraction, reward; and instinct of fears (appeal to fear), such as illness, weaknesses, loneliness, need, uncertainty, security, prejudices, learned opinions and comforts. In this sense, 'all human needs, relationships, and fears – the deepest recesses of the human psyche – become mere means for the expansion of the commodity universe under the force of modern marketing' (McChesney, 2008).

2.1.4.3 Politics

Advertising can also have impact on politics. Over 650 million euros was spent for the Barack Obama campaign, alone, on advertisements and TV commercials in the election for the US presidency in 2008 (Luo, 2008). Basically, it gives the politics the chance to respond charges very quickly, reaching a few million viewers. But since this is very expensive, only very rich people (or parties that can get huge amounts of money) have the chance to run for a political position, or at least depend on the donation of wealthier people who could have a huge impact on democracy these days. The political issues talked about in an election are also very much simplified because the spots are only about 30 s long, and you cannot discuss much in such time period.

2.1.4.4 Culture

Finally, it impacts the culture of a country. By living in a global economy, commercials are often produced once and then exhibited in several different countries, which leads to break down the differences of these societies. Children will grow up without knowing how the culture was earlier in their country. It can also lead to a lot of discussion about moral values if we just think about some controversial ads, which are present in our daily lives. Although the 'hyper-commercialization' of the culture is recognized and roundly detested by the citizenry, the topic scarcely received a whiff of attention in the media or political culture. Nevertheless, with the current regulation organisms present in most societies, these problems are usually minimized.

Therefore, more than ever, it is necessary to re-think the way advertising will be deployed in the future, as it will affect different environments in a myriad of ways. A good example for such premise is depicted in Figure 2.5, which shows the way Internet sees advertising (the keywords were generated using the Personas tool from MIT Media Lab, http://personas.media.mit.edu).

2.1.4.5 The players

The previous sections presented how advertising has evolved over time, in which ways it can take form and what are its collateral effects. However, they did not

Figure 2.5 Advertising related fields

mention the differing entities involved in the complex value chain that is advertising. This will be particularly important, when later, in section 4.3.3, the different implementation alternatives for the proposed advertising solution are depicted.

Therefore, to put them into context, one must first understand who the stakeholders are. Advertising is a complex business with considerable local variation. For this reason, traditional value chains do not accurately describe the market. Instead, a value web is used to depict the stakeholders and their relationship with one another (see Figure 2.6).

The driving force begins with an *advertiser's* desire to convey a message to a group of *consumers* (or users) in order to increase brand awareness and generate sales. Working together with a *media agency*, the advertiser defines a marketing mix and branding objective. The media agency identifies the channels and media mix that best suit the campaign. It is also in charge of buying inventory from *sales houses* and *publishers*, and follows up the results of the campaign. The term inventory is used to denote the 'real estate' that is available for the display of advertising within the telecom and Information Technology (IT) domains. Examples include SMS, MMS, the 'idle' screen of a mobile phone, ring-back, and pre- and post-rolls in IP Television (IPTV), radio, computer and mobile TV sessions. Sales houses are players that partner with publishers to resell advertising space on their behalf. Because advertising is traditionally a funding element, as opposed to a stand-alone service, the attractiveness of the content or the context in which it is displayed is very important – this is the inventory that publishers provide. *Operators* assume the role of publisher whenever advertisements are placed or sold in

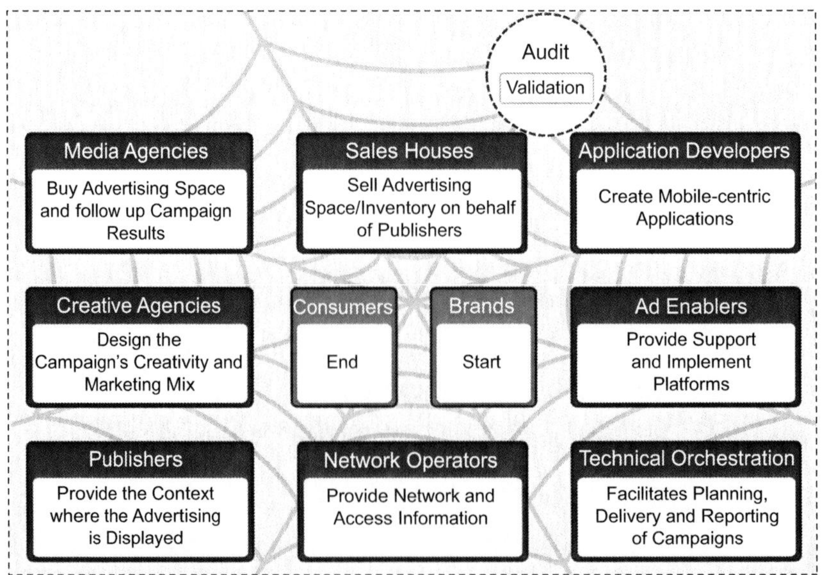

Figure 2.6 Players in the advertising value web [Adapted from Asp et al. (2009)]

telecom delivery channels, such as SMS, MMS, WAP portals, IPTV/video, mobile TV and ring-back tones (Asp et al., 2009).

2.2 Next generation networks and services

From the previous section, it is possible to understand that advertising involves several players across a fairly complex value chain. Furthermore, it showed that due to the myriad of alternatives to consume advertisements, the number of devices and infrastructures needed to support the aforementioned needs to be carefully considered. Nevertheless, as this work focuses on real-time and dynamic multimedia-based advertising, it is important to understand the underlying concepts and technologies beneath the service layer. In this sense, this section covers the evolution towards all-IP networks and details some of the standardization efforts to enhance next generation networks and services.

2.2.1 Convergence: evolution towards all-IP

The telecommunication industry faces a major evolutionary step for the next years. In a recent past, pure Public Switched Telephone Network (PSTN)-based communication services have become less predominant, whereas competitions for new revenues are forcing the companies to transform their existing business models. Therefore, the tremendously evolving Internet, as well as the use of IP-based transport networks, presents a natural environment to enable the provisioning of services, independently of the underlying networks. This led to the term of 'convergence', which refers to separate technologies that can share resources and interact with each other, collaboratively, aiming at creating new efficiencies (Figure 2.7).

Although the term convergence can be applied to several distinct concepts throughout this work, as seen in Figure 2.8, this section gives particular emphasis to Fixed Mobile Convergence (FMC). In general, FMC usually specifies the bundling of fixed and wireless telecommunication networks, to deliver a seamless and similar user experience that can be consumed anywhere, at any time and

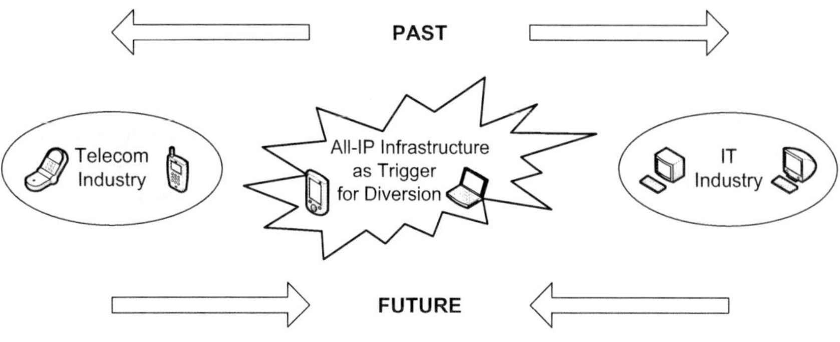

Figure 2.7 IP as trigger for diversion

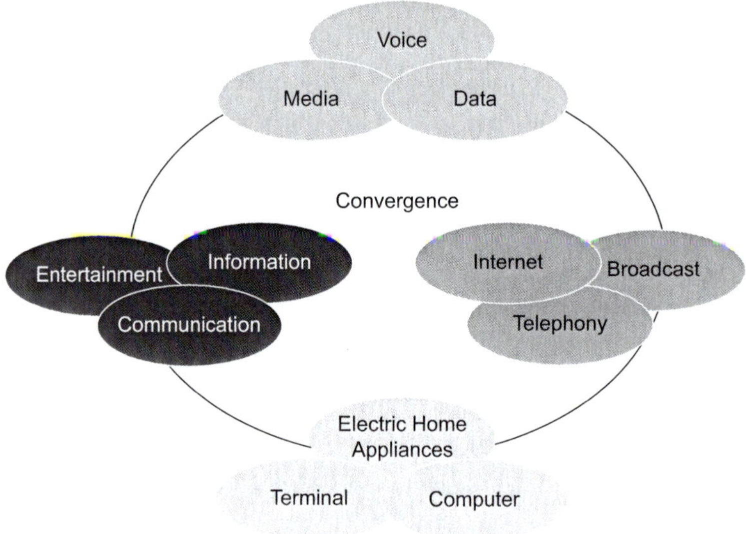

Figure 2.8 Various characteristics of convergence (Eskedal, 2008)

irrespective of the network, access technology or device being used. This ranges from using a single device which interconnects wired and mobile networks to the switches between these networks that allow using a mobile service at the expense of fixed activity. In terms of Next Generation Networks (NGNs), the main aspects of FMC relate to device, service and network convergence.

2.2.2 *What is a next generation network?*

With the purpose of providing a common architecture to enable different multimedia services across heterogeneous networks, the International Telecommunication Union, Telecommunication Standardization Sector (ITU-T) has introduced the concept of NGN, under the following recommendation (2006). It was defined as

> a packet-based network able to provide telecommunication services and able to make use of multiple broadband, QoS-enabled transport technologies and in which service-related functions are independent from underlying transport-related technologies. It enables unfettered access for users to networks and to competing service providers and/or services of their choice. It supports generalized mobility which will allow consistent and ubiquitous provision of services to users.

As a consequence, a formal NGN architecture representation is depicted in Figure 2.9. It highlights a horizontally layered network architecture, instead of the common vertically separated network for each service, providing the advantage of services to be offered separately and to evolve independently. It uses the IP-based

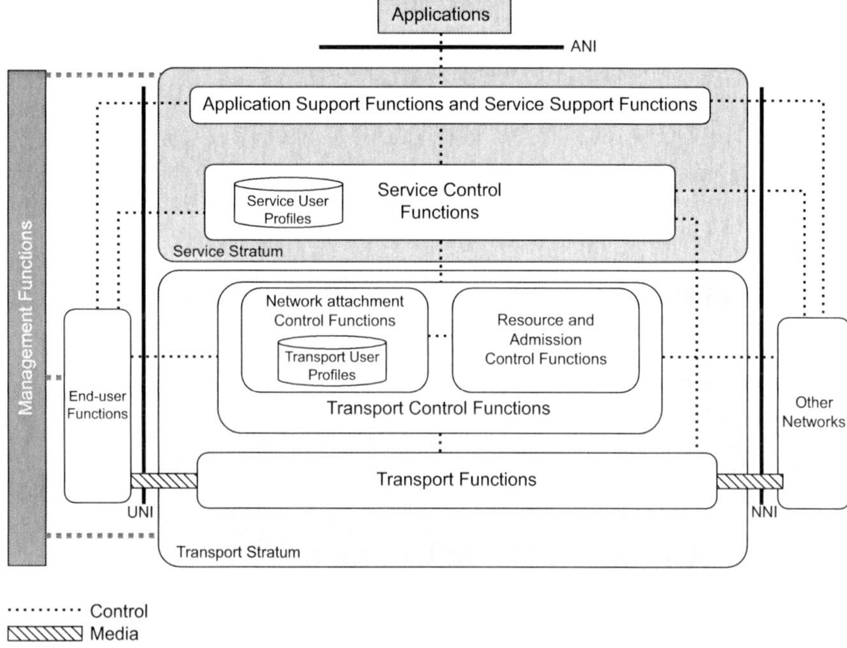

Figure 2.9 Formal NGN architecture overview (ITU-T Rec. Y. 2012)

transport for all services and defines the formal components' reference points, namely the Network–Network Interface (NNI), the User–Network Interface (UNI) and the Application–Network Interface (ANI).

2.2.3 Service delivery for next generation networks

The emerging shift from circuit-switched networks towards packet-switched core networks has led to unified all-IP network infrastructures, capable of providing FMC and triple-play, as well as prospective (multimedia) services. In this sense, international standardization efforts around NGNs started to appear. The most relevant was probably the Third Generation Partnership Project (3GPP) Universal Mobile Telecommunications System (UMTS) Release 5, as it marked the beginning of studies on a next generation telecommunications infrastructure for high-speed broadband access, with focus on mobile networks. Labelled as IP Multimedia Subsystem (IMS) (IMS, 2006), it has evolved since then as a general NGN control overlay network on top of any mobile, fixed and even cable networks. Furthermore, mostly due to the joint efforts with other standardization bodies, such as 3GPP2, European Telecommunications Standards Institute (ETSI) Workgroup Telecoms & Internet converged Services & Protocols for Advanced Networks (TISPAN) and CableLabs, it has been constantly enhanced in subsequent releases. Currently, the industry is working on release 10 and the architecture covers most of the existing access networks (e.g. Long-Term

Evolution (LTE), UMTS, Worldwide Interoperability for Microwave Access (WiMAX), Wireless Local Area Network (LAN) (WLAN), as well as traditional legacy systems).

Like NGNs, the IMS uses a layered architecture, which consists of three planes: user, control and application. In addition to NGNs, the transport layer of the ITU's approach is sub-divided. The transport and bearer services (user plane) are separated from the session signalling services, representing the control plane. The key feature of the layered approach is the extension of the IP network, from user endpoint to the service or called party, while remaining independent of the type of access network. Thus, it works both with legacy networks and with additional new access networks. The application plane hosts application and content servers capable of executing and

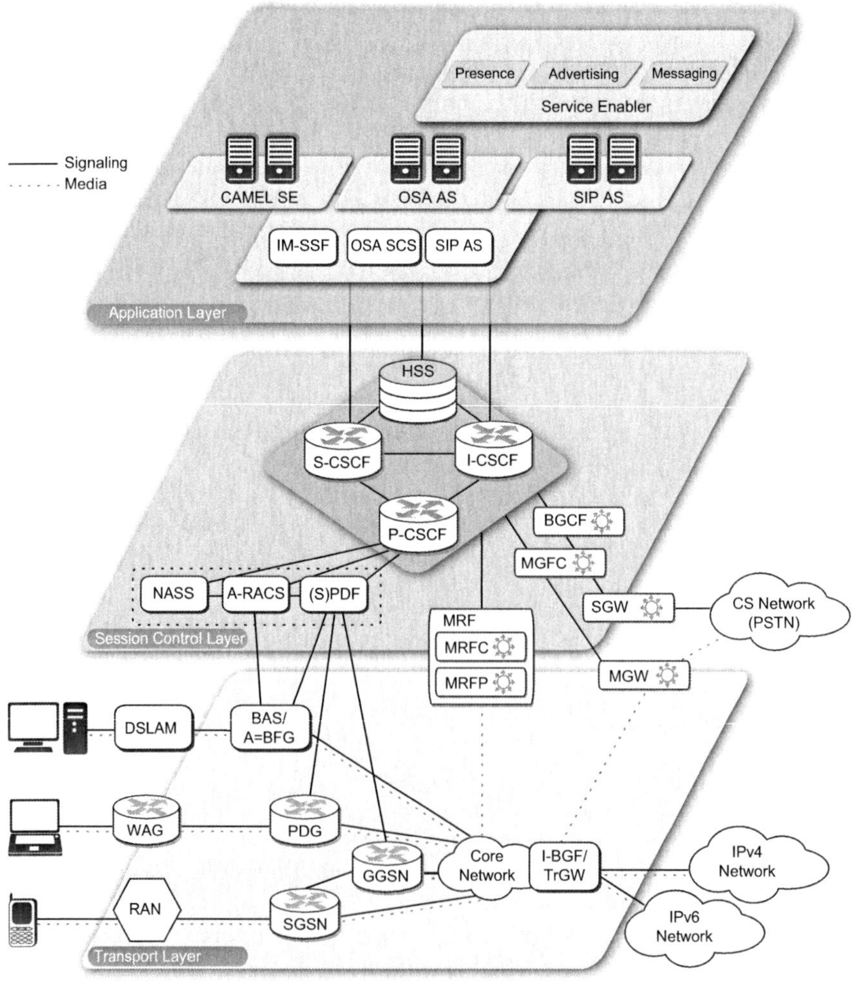

Figure 2.10 IMS system architecture overview

proving value-added services. User- and service-specific information are additionally situated within this block. The control plane, or rather, the IMS layer, comprises components that are responsible for call and session control, such as creation, modification and release. The user/transport plane describes the connectivity layer that implies gateways, routers and switches for both the backbone and the access networks. Figure 2.10 depicts the simplified IMS architecture.

From an architectural point of view, 3GPP defines an IMS core layer comprising a set of extended Session Initiation Protocol (SIP) servers, called Call Session Control Functions (CSCFs), as well as signalling media gateways to interwork with fixed and mobile legacy circuit-switched networks and media servers. End-users will use an enhanced SIP/IMS client to connect to the system. A specialized Authentication, Authorization and Accounting (AAA) server known as Home Subscriber Server (HSS), or in case of fixed network access, referred to as User-Profile Server Function (UPSF) by ETSI TISPAN, stores the entire user- and service-related data, which is exchanged with the SIP servers using the Diameter Base protocol. Application Servers (ASs) are out of scope from the 3GPP perspective, but IMS provides standardized interfaces based on SIP and Diameter for the application layer, as a single point of integration for service platforms.

2.2.4 *Service execution and exposure environments*

In order to accelerate service deployment and time-to-market, while achieving an optimum cost structure, the industry has adopted Service Oriented Architecture design principles in terms of Service Delivery Platforms to minimize the demand for integration. Therefore, the Session Description Protocol (SDP) establishes a set of abstract interfaces that protect applications from the complexity of the underlying network infrastructure (e.g. for service interconnection) and further integrates support and management applications, such as Customer Relationship Management (CRM), Business Support System (BSS)/Operating Support System (OSS) and service activation. Additionally, the SDP controls access to network resources and performs interfacing with external applications and third-party service providers, to facilitate secure access to network resources and service enablers (Maes, 2008). Using different terminologies, but providing a common way of interfacing, the following subsections present some of the most common approaches to address service execution and exposure.

2.2.4.1 Application servers

As seen from the service perspective, the core components for building services on top of an IMS infrastructure are the CSCFs cooperating with the service domain via the IMS Service Control (ISC) interface that forwards SIP requests towards ASs. An AS provides the service logic and service creation environment for applications and services, whereas the server resides in the user's home network, or in a third-party location. Therefore, it can act as an IMS service element on behalf of native application/dedicated platforms, as well as an external service, or service interworking element, on behalf of gateways to legacy service environments/frameworks.

Although there are different types of ASs, this work will focus on SIP ASs. They run natively in IMS, hosting conventional services based on SIP and are intended to be the default ASs for developing and deploying new services, by supporting a variety of well-standardized development Application Programming Interfaces (APIs) and technologies. Depending on the application, they can operate in three different modes (OEC, 2008):

- *SIP proxy mode*: listens to traffic. These can be used for logging or simple manipulation.
- *SIP User Agent (UA) mode*: acting in terms of an originating/terminating UA (e.g. endpoint) or a redirect proxy.
- *SIP Back-to-Back User Agent (B2BUA) mode*: acting as a User Agent Server (UAS) in the originating call leg and as a User Agent Client (UAC) in the terminating call leg, thus concatenating two SIP end-points (e.g. click-to-dial).

In order to create services, the SIP ASs support both low-level and high-level APIs, where several solutions have been devised in the course of time. Low-level APIs are running directly on the server, in contrary to high-level APIs, which make use of middleware technologies.

The first group includes scripting languages like SIP Common Gateway Interface (CGI) (Lennox et al., 2001) and Call Processing Language (CPL) (Lennox et al., 2004). SIP stacks for different programming languages as well as proprietary APIs. But in the majority of the cases, current solutions leverage the Java programming language, employing either SIP Servlets or Java API for Integrated Networks (JAIN) APIs. SIP Servlets are specified through the SIP Servlet API (Java Community process, 2008). It adopts the already existing Hypertext Transfer Protocol (HTTP) Servlet architecture and also defines mechanisms to develop converged SIP-HTTP applications. JAIN has been developed to meet the requirements of telecommunication services in tier-one service providers, with special focus on high throughput and low latency.

2.2.4.2 Service enablers

Alongside with IMS, the concept of independent service building blocks has emerged as 'service enablers'. These key elements represent generic and reusable building blocks for service creation. According to web commentators (Barton et al., 2000), the enablers constitute a logical layer and the service enabling layer, which resides between the application layer and the control layer. In fact, the term has been mainly associated with Open Mobile Advertising (OMA), and the paradigm of modular and reusable elements was initially born during the specification of a Push-to-talk over Cellular (PoC) service that allows a number of users to communicate in a walkie-talkie way. (See the Open Mobile Alliance push-to talk over cellular at http://www.openmobilealliance.org.) PoC uses presence, group management and Instant Messaging (IM) capabilities to provide information to the users, as well as to the service itself. Afterwards, many enablers with different capabilities have been specified.

The definitions of several application service enablers by the OMA and the need for a general access function for third-party service access led to the specification of the OMA Service Environment (OSE). It provides an architectural framework that describes the interaction with and among the OMA enablers, services and resources. The OSE domain incorporates specific enabler components and follows the aspects of controlling the access to services by using policies. Figure 2.11 depicts the OSE architecture.

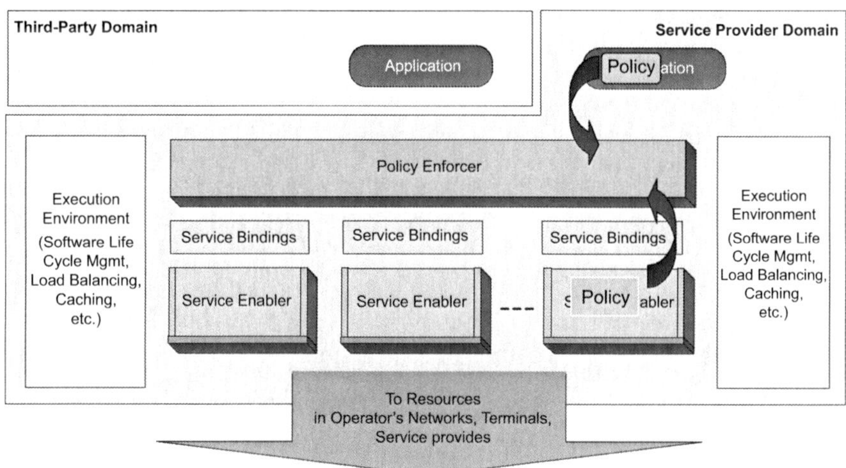

Figure 2.11 OMA service environment architecture

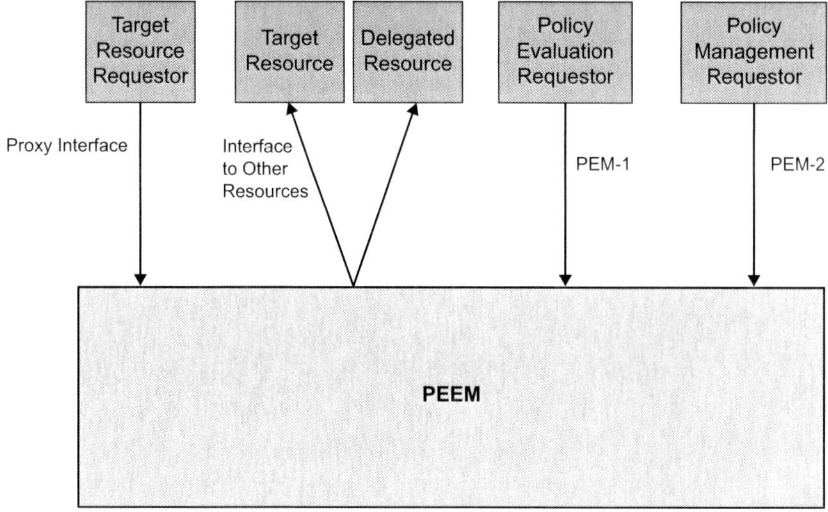

Figure 2.12 OMA Policy evaluation, enforcement and management architecture

Policies are loaded dynamically for policy evaluation and enforcement via a policy enforcer. This policy enforcer, which has been introduced by the OMA as Policy Evaluation, Enforcement and Management (PEEM), intercepts service requests from any domain, or requestors and apply certain rules (policies) on them. Policies in this context are formalisms used to express business, engineering or process criteria, represented by a combination of policy conditions and actions. Figure 2.12 provides an overview of the PEEM architecture (by OMA), plus the logical components, interfaces and relationships the PEEM has with other entities in the OSE. Note that the PEEM enabler implementation can evaluate and enforce policies when exposing any resource (e.g. application, enabler, component, etc.). The PEEM enabler exposes the following interfaces:

- *PEM-1*: PEEM-specified callable interface.
- *PEM-2*: PEEM-specified management interface.
- *Proxy interface*: used for intercepting requests to target resources.

In addition to the PEEM components and interfaces, there are other elements represented in Figure 2.12, which can be shortly described as:

- *Target resource requestor:* a resource (e.g. application, enabler) that issues a request to a target resource.
- *Target resource*: the destination resource for a request made by another resource.
- *Delegated resource*: a resource to which the PEEM may delegate certain policy actions during the policy processing process.
- *Evaluation requestor*: resource that issues a request for policy processing by the PEEM. An evaluation requestor uses the PEM-1 interface.
- *Management requestor*: a resource that issues a request for policy management to the PEEM.

And finally, PEEM supports two options for expressing policies:

- *A rule set-based option:* each rule is evaluated as separate entity, and the combination of the results of processing all the rules in the rule set determines the policy outcome (notice that a precedence mechanism may be needed). The policy expression language specified is the Extensible Markup Language (XML) Common Policy schema from Internet Engineering Task Force (IETF) (Schulzrinne et al., 2007).
- *A workflow-based option*: the entire policy is processed as a whole, following a flowchart approach, where at each node in the graph, a rule is being processed. The policy expression language specified is Web Services Business Process Execution Language (WSBPEL) defined by the OASIS consortium (Oasis WSBPEL, Version 2.0, April 2007).

In this sense, a PEEM function forms the main integral component of an OSE and provides additional functionality based on the definitions of policies for the OMA enabler concept.

2.2.4.3 Other API-based approaches

Although the term API has a long history in the IT world, the same started to be associated with telecommunications around 1998. Since then, these APIs allow flexible implementation of services in ASs accessing the network functions, via dedicated interface operations defined by the API. They are independent of the underlying network signalling protocols through the definition of protocol-specific resource adaptors and should thereby support network and vendor independence. Service independence is provided by the extensibility of an API. The major target of an API approach, however, is when publishing the API specifications openly, to allow the creation of services by a third-party community and thereby gaining more service innovation. Despite this we will not cover all the initiatives, below it is possible to find the most relevant efforts and respective references for further details:

- *Open Service Access (OSA)/Parlay (http://docbox.etsi.org//tispan/open/osa/Overview.html)*: defines an open service architecture that enables IT applications to access Telco features through a standardized API, which maps the network's resources into a set of interfaces, called Service Capability Functions (SCFs).
- *Parlay X (http://docbox.etsi.org//TISPAN/Open/OSA/ParlayX30.html)*: created to facilitate usability, it deliberately comprises a much abstracted set of APIs that can be mapped to the offered SCFs, or directly to the network functionalities. Consequently, the applications can access those functions through web services' APIs and be developed in any programming languages that support web services (3gpp ts 29.199-xx v8.0.0, OSA; parlay x web services, September 2009).
- *Service Capability Interaction Manager (SCIM)*: defined by 3GPP, it represents a function, which can be considered as a service broker for SIP services, acting within the SIP application layer of IMS, managing interactions between ASs (3gpp ts 23.002 v5.12.0, 3GPP; technical specification group services and systems aspects; network architecture (release 5), September 2003).
- *Java Community Process (JCP)*: adopts the idea of a SCIM, as a so called Application Router (AR) as part of its SIP Servlet 1.1 specification JSR 289 (Java Community process, 2008). The AR is defined as a function alongside the Servlet container. Then, it is responsible for routing SIP requests to applications.
- *TeleManagement Forum (TMF, http://www.tmforum.org)*: an industry association that is working on business-oriented policies and solutions. It defines two approaches. IPsphere (http://www.tmforum.org/ipsphere) represents a SOA-based framework for enabling service provider collaboration and resource federation. The TMF Service Delivery Framework (SDF) (http://www.tmforum.org/ServiceDeliveryFramework) outlines a service delivery mechanism that transcends multiple SDPs and domains with increased emphasis on managing the full service lifecycle.

2.2.5 Multimedia delivery and related protocols

Multimedia content delivery is dependent on a myriad of factors. It depends on the network type, architectural model of the service, transmission mode, content type, protocols used, etc. Nevertheless, with the convergence over IP networks, some of trends started to unify even in distinct environments. Therefore, this subsection will overview some of these technologies.

2.2.5.1 Transmission modes

The transmission mode relates to the way multimedia flows from the origin until the destination. In the IT terminology, it is often referred to as routing mechanism as well. Independently of a number of originators or destinations, there are three main transmission types:

- *Unicast*: is associated with the transmission of a single source to a single destination. Due to its capability to support bidirectional communications, it is widely used in web and mobile application domains.
- *Broadcast*: refers to a unidirectional method of transferring a message to all recipients simultaneously, where a packet transmitted will be received by every device on the network. In practice, the scope of the broadcast is limited to a broadcast domain. It is commonly used for content delivery on access networks such as the Digital Video Broadcasting (DVB) or 3GPP Multimedia Broadcast Multicast Service (MBMS).
- *Multicast*: represents the delivery of a message or information to a group of nodes simultaneously, in a single transmission from the source, creating copies automatically in other elements, only when the topology requires it. In short, the content is transmitted from a single source to multiple destinations. However, this type of transmission can be realized in different ways:
  (a) *Access layer*: the routing occurs at layer 2 (from the Open Systems Interconnection (OSI) model), and it means all nodes (e.g. routers, switches, hubs) in the network must support it. As an example, DVB, MBMS or WiMAX supports it.
  (b) *IP core*: the key concepts include an IP multicast group address, a multicast distribution tree and receiver-driven tree creation. In this method, routers actively participate in multicast, making copies of packets as needed and forwarding towards the multicast receiver. Like in (a), users must join this address to start receiving content.
  (c) *Application layer*: replicates the multicast concept at the application layer, where end-systems communicate through an overlay structure, assuming unicast paths (and possibly multicast for overlay networks) provided by the underlying network. This variant is not so efficient like the previous one (due to packets' replication); nevertheless it is much more flexible and can be implemented across heterogeneous networks.

In this sense, depending on the application, the network structure, or the type of multimedia, one of the aforementioned methods might be the most appropriate.

2.2.5.2 The protocols

In an NGN environment, multimedia content delivery protocols are primarily based on Internet protocols defined by the IETF. Although it is not possible to cover all of them, this subsection will introduce the most relevant for this work.

Session initiation protocol

Session Initiation Protocol (SIP) has become a strong, catalytic force shaping today's telecom industry. Defined by the IETF under the RFC 3261 (Rosenberg et al., 2002), this protocol represents a key ingredient in the converging world of telecommunication-based applications. It is an application layer protocol for controlling and managing any multimedia session. It can be used for creating, modifying and terminating two-party (unicast) or multi-party (multicast) sessions consisting of one or several media streams. The modification can involve changing addresses or ports, inviting more participants and adding or deleting media streams. SIP employs design elements similar to the HTTP request/response transaction model. Each transaction consists of a client request that invokes a particular method or function on the server and at least one response. SIP reuses most of the header fields, encoding rules and status codes of HTTP (e.g. Uniform Resource Identifiers (URIs) and many message header fields, such as 'To' and 'From'), providing a readable text-based format.

According to the IP paradigm, SIP is an end-to-end signalling protocol, following the session-based control model, in which, an SIP entity may act as a client and as a server concurrently. Therefore, it is used to control sessions, rather than the transmission of media traffic (e.g. voice packets). Description and transport of content are managed by other protocols, which are commonly used in conjunction with SIP, such as the SDP or Real-Time Transport Protocol (RTP), respectively.

In addition, SIP defines four types of logical entities and each may act as: a client by initiating requests, a server by responding to requests, or both. These are: UA, SIP proxy, registrar and SIP redirect server. As for the methods supported, there are six main commands: Invite, Ack, Bye, Cancel, Options and Register. Finally, it is highly extensible, and multiple RFCs and other documents define new extensions (e.g. additional methods) to the SIP specification (Rosenberg, 2008).

Session description protocol

Session Description Protocol (SDP) is a format for describing multimedia sessions' parameters. Defined by the IETF in RFC 4566 (session description protocol, IETF, June 2002), it is intended for the purposes of session announcement, session invitation and parameter negotiation. Although SDP does not deliver media itself, instead it is used for negotiating the media type, format and all associated properties between end-points. It follows the offer–answer model, where for each received request (offer), a response (answer) must be created. The set of properties and parameters are often called a session profile. Furthermore, it is designed to be extensible to support new media types and formats. It is usually used together with SIP and RTP.

Real-time transport protocol

Real-Time Transport Protocol (RTP) defines a standardized packet format for delivering multimedia (e.g. audio or video) over IP networks. It was specified by the IETF in RFC 3550 (Schulzrinne & Casner, 2003) to be a network- and transport-independent protocol. Furthermore, it was designed for end-to-end, real-time, transfer of data streams and provides facility for jitter compensation and detection of out-of-sequence arrival in data, which are common problems during transmissions on an IP network. RTP supports data transfer to multiple destinations through multicast.

Since there is no guaranty (assuming User Datagram Protocol (UDP) is used) that RTP data packets arrive in order they were sent, packet reconstruction needs to be done at the application level, by using the information provided in the packet header. While RTP does not provide any mechanisms to get feedback on quality of data transmission and information about participants in an on-going session, these issues are addressed by control protocols like Real-Time Control Protocol (RTCP) (its basic functionality and packet structure are defined in the RTP specification).

Real-time streaming protocol

Real-Time Streaming Protocol (RTSP) is a network control protocol designed for use in entertainment and communications systems, to control streaming media servers. Defined by the IETF in RFC2326 (Schulzrinne et al., 1998), it is used for establishing and controlling media sessions between end-points (client and server). It provides remote control functionality for audio and video streams, such as play, pause, fast-forward, reverse and absolute positioning, where the data source can be live or stored content. The transmission of streaming data itself is not a task of the RTSP protocol; therefore, it is usually used in conjunction with RTP or SDP and provides an alternative to SIP for some applications.

2.2.6 Summary

Although, at a first glance, it might not seem to be directly related with advertising technologies, the role of NGNs is preponderant for the development of future advertising initiatives. In fact, these will provide the foundation for any type of real-time multimedia-based advertising. To assure a fast, scalable and interoperable environment, it is important that all concepts developed take standardization efforts under consideration, otherwise they might become isolated in the long term. In this sense, this section overviewed the state of the art for NGNs and showed some of the service creation principles that should be adopted by any multimedia service in general. In addition, it presented some of the most common protocols used to manage multimedia content over IP-based networks. Altogether, using the aforementioned architectures, components, concepts and protocols will allow services to be compliant with major telecom and IT-related guidelines.

2.3 Advertising state of the art

Due to the variety of existing advertising types and channels, like introduced in section 2.1.3, this work will only focus on solutions, services or relevant tools that cover either personalized (the user) or contextualized (specific group of users) advertising within real-time systems. Working within this field, there are several initiatives that come from different backgrounds, namely, standardization and industry and research entities. While the first tries to regulate and propose standardized formats, metrics, processes and interfaces, the second focus on the deployment of these same ideas into real-world products. Research on the other hand covers either particular technological issues that can be then user-cased as advertising or on a socio-psychological level, dealing with the challenges and factors that might influence advertising within next generation multimedia services.

2.3.1 Standardization

2.3.1.1 Advertising

The hype around advertising as the business model for the next generation services raised significant interest not only for advertisers and media publishers but also for standardization bodies, which mainly comprise network operators, device and network suppliers, IT companies and content providers. Although these cover most technologies such as television and Internet, the mobile world is where this work is most prominent. As businesses begin to recognize the potential of mobile advertising, the need for specific guidelines in this area is becoming starkly apparent: standards are needed to gain the confidence of advertisers and ad agencies to establish the mobile handset as a viable alternative media channel. It is necessary to define a common understanding of what will be delivered to users, which is consistent, regardless of which operator/provider is managing the network. It is vital that the value chain for all players involved be clearly defined and expectations or inter-relationships between players are fully understood, in order for the business to flow freely and the lofty ambitions for this industry be fully realized.

Among these, the most notable is the OMA (www.openmobilealliance.org), whose main goal is to facilitate a global adoption of mobile services and ensure service interoperability. It has just completed a key phase in defining the architecture on which this technology will run. Looking mainly at the technology aspect of mobile advertising, it addresses the major challenges of interoperability between devices, geographies, platforms, service providers and applications – these go from personalization, interactivity, metrics to ease reporting. The first two of these embrace broadcast services such as TV and other content distribution services and cover issues like location, presence, user profile, device type and configuration, as well as Peer-to-Peer (P2P) services, browsing and MMS, where technologies in the devices need to facilitate features like Click2call, Click2buy, voting and coupons. The OMA also specifies guidelines regarding advertising metadata, its description and usage and the enhancement of user profiles with advertising preferences and formats, to improve the user experience and interoperability. The end product of these deliberations – the MobAd Enabler – will provide a consistent architecture so that service providers can send personalized and interactive advertisements, as well as give them the ability to reliably and consistently measure mobile ad effectiveness, across the range of different types of adverts, independently of the format.

Focusing on the user experience, the Mobile Marketing Association (MMA) has defined the '6 C' code of conduct for mobile advertising players: choice, control, constraint, customization, consideration and confidentiality. Furthermore, they specified a set of mobile advertising guidelines, which provide recommendations for the global ad units, broadly used in mobile advertising in the following mobile media channels: mobile web, messaging, applications, mobile video and TV. The Guidelines recommend ad unit usage best practices, creative technical specifications, as well as guidance on ad insertion and delivery, necessary to implement mobile advertising initiatives. The guidelines are intended to promote the development of advertising on mobile phones by:

- Reducing the effort required to produce creative material.
- Ensuring that advertisements display effectively on the majority of mobile phones.
- Providing an engaging, non-intrusive consumer experience.

On the other hand, the Global System for Mobile Communications Association (GSMA) is facilitating a range of cross-operator initiatives to accelerate the development of mobile advertising, and to build trust and confidence in mobile as a medium. As the market for mobile advertising develops, the media industry is demanding that an independent and robust approach to mobile audience measurement is developed. In this sense, through its Mobile Advertising Program, a feasibility study is being conducted to explore the aggregation of valuable operator data on an independent, consistent and audited basis, to deliver cross-operator metrics to the media and advertising communities. The group consults the advertising industry to secure its support for metrics that will form a common mobile media currency. The combination of a widely adopted common currency with independently aggregated and audited census-level data represents a chance for mobile to differentiate itself from all other media in terms of the granularity and transparency of its reporting. For traditional media, audience measurement is commonplace (e.g. listener/viewer/reader numbers). For the Internet, audience measures, such as unique user and page impression have been developed to enable media owners to quantify the performance of their media properties, enabling media planners to understand consumer consumption of the Internet, and brands or advertisers to assess how to communicate with their customers.

Significant work has been also achieved by the Mobile Entertainment Forum (MEF), the Interactive Advertising Bureau (IAB) and the World Wide Web Consortium (W3C), where the evolution of the mobile entertaining industry, the effectiveness of interactive advertising and the issues related with the mobile production chain are addressed, respectively.

2.3.1.2 Summary

Major standardization initiatives come from a set of players that want to get a piece of the share in the profitable world of advertising. This is most notable in the mobile industry, where players are still trying to get market quota. That is why manufacturers, telecommunication and service providers are working together to

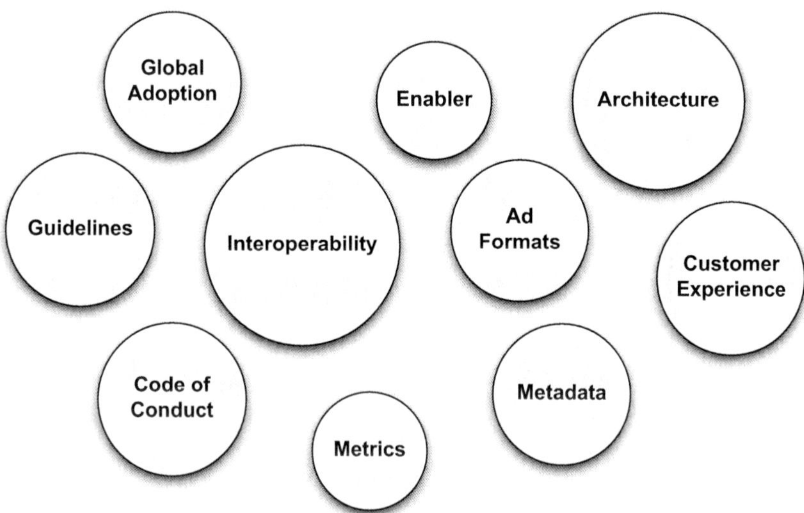

Figure 2.13 List of topics being addressed by standardization bodies

achieve the necessary interoperability that will allow a minimum market penetration, a 'mandatory' requisite to enter the 'game'. Altogether, the work from standardization bodies shall help the industry to develop, but should not limit the creativity for innovation. Therefore, it is recommendable that most solutions follow the aforementioned guidelines. However, it is also expected that some proprietary technologies or ideas might come out in parallel to offer a differentiated value towards the end-user (consumer). To help understanding the major trends within the standardization bodies, Figure 2.13 resumes some of the most dealt topics among them.

2.3.2 Research and academia

From the research point of view, this topic is also generating several contributions. Exploring past experiences and looking at the future, the work produced by Ranganathan and Campbell (2002) and Eriksson and Akesson (2008), respectively, helped to identify what are the main challenges ubiquitous advertising will be facing in the future:

1. Reaching the right people with the right ads.
2. Delivering ads at the right time.
3. Delivering ads in the right way.
4. Serendipitous advertising.
5. Providing means for the users to follow up the ad.
6. Collecting advertising revenues.
7. Privacy and security.

2.3.2.1 Context-awareness

Looking at the previous enumerated list, it seems that most of the points have a common determiner, the user. In this sense, it became a focal point to know more about them. The first steps towards this direction were given by Weiss et al. (2008) and Hsieh et al. (2008), which proposed extending user basic profile and preferences' data to accommodate other sources of context information. This was further explored by Liapis et al. (2008) and Sanchez et al. (2008), where location information was being used together with user profile data to empower users with a more compelling and interesting advertising experience by using Global Positioning System (GPS), GSM, wireless networks, bluetooth, Radio Frequency Identification (RFID) and other location sensor-based technologies. Unfortunately, the attempt of combining different sources of context information to improve the targeting process is not very well explored.

On the other hand, sometimes, instead of contextualizing the user, it is more interesting to understand the situation or the circumstances under which interactions occur. This represents an alternative way of profiling and is particularly useful for real-time communications. This is what Google does inside their own Google Mail accounts, providing the user with advertisements, which are related to the content of the e-mail messages. Using a similar approach, the work done by Brenner and Wang (2009) and Zadeh and Moshkenani (2008) proposes mining the bit pipes, trying to discover and leverage users' behaviour. By doing so, they allow real-time content analysis, prevenient from IM, e-mail exchanges or other social communication platforms, and support keyword matching, syntax, semantic and statistical analysis. In these steps, users are categorized based on the topic exchanges by them in the network. Considering that these topics show the user will, finally relevant advertisements are represented to them. A similar approach is taken by Nagarajan et al. (2009) and presents a system that enables generation of profile ads from user activity by assessing the monetization potential of the activity and then identifying relevant keywords for advertising.

2.3.2.2 Social networks

Still related with social networks, several researchers try to explore the links people establish between themselves. Despite the purposes may vary, the common principles remain the same, explore the relationship between individuals. The approach in Yang et al. (2006) starts with the discovery of cohesive sub-groups (clusters) from social networks. Then, the discovered cohesive sub-groups are used to identify a shortlist of prospective customers for a given product. Trying to investigate the degree to which social network information complements existing user-profile information for targeting, Bagherjeiran and Parekh (2008) found that there is significant evidence in our social network of homophyly that links in the network indicate similar ad-relevant interests. Therefore, they proposed an ensemble classifier to combine existing user-only models with social network features to improve response predictions. The results showed that the response rate on ads is proportional to the number of friends who have responded to an ad in the past.

Focusing specifically on the type of links between peers inside social networks, Lin et al. (2009) proposes the use of the network structure to derive which actors in the social network site are more influential, then using word of mouth thread to fulfil the advertising recommendation. Similarly, inside a more realistic and ubiquitous scenario, the research in Lee et al. (2009) introduces the concept of Connected Consumption Network (CCN) that allows a community of consumers to collaboratively sense the market from a mobile device, enabling more informed financial decisions in geo-local context. In other words, by building a system of actual transactions and establishing connection to share these purchasing experiences among friends and social network, the quality of information and recommendations can be highly increased (through social filtering). This opens the path for transforming undesired advertising into a new paradigm called recommendations.

2.3.2.3 Ubiquity

Due to the ubiquity of the world we are living today, it is expected that advertising experiences some of these consequences. At the same time, ubiquitous computing applications will very likely be supported by advertising, continuing the success of advertising on the Worldwide Web and paralleling the predicted growth of mobile advertising. More interesting, however, is that ubiquitous computing will eventually support advertising in several ways. The survey work done by Krumm (2010) explains how advertisers are already adopting certain ubiquitous computing technologies, and show how ubiquitous computing research can help advertisers in the areas of ad targeting, ad feedback, customer awareness and privacy. This reinforces the aforementioned importance of context-aware advertising systems, which should be capable to react and adapt to context changes.

Nevertheless, most of the times, even in the possession of several user-related data, it is necessary to compare it with other source to improve the relevance of the predicted outcome. In fact, information-retrieval systems conventionally assess document relevance using the bag of words (context) model. Consequently, relevance scores of documents retrieved for different queries are often difficult to compare, as they are computed on different (or even disjointed) sets of textual features. Many tasks, such as federation of search results or global thresholding of relevance scores, require that scores be globally comparable. To achieve this aim, Agarwal et al. (2009) propose methods for non-monotonic transformation of relevance scores into probabilities for a contextual advertising selection engine that uses a vector space model, where the calibration of the raw scores is based on historical click data. This helps improving the accuracy/targeting efficiency of the provided advertisements.

2.3.2.4 Privacy

While the benefits of personalization, contextualization and adaptation within advertising scenarios are obvious, the privacy and security concerns associated with it are extremely high. In a survey done by Kern et al. (2008), through structured interviews with potential users in two European countries, the authors explore the reactions to the ideas of implicit and explicit generations of personal user profiles and relate these to more general user views on privacy and targeted advertising.

The initial results suggest that users are prepared to trade off ease of use against increased levels of control over their data and are therefore more comfortable with an explicit system. In other words, they prefer a web-based, decoupled approach that allows them to access and change their profile randomly. The fact that fully-implicit profile capture was rated as being by far the easiest system to use, coupled with the comments received while conducting the interviews, further suggests that users are indeed aware of the additional effort such a scheme imposes on them when compared to fully-implicit profile capture, but are – at least in theory – willing to trade this inconvenience off against the ability to access and modify their profile at their liking.

Another study performed by Yang and Wang (2009) examines the effects of information sensitivity and compensation on privacy concern and behavioural intention, where behavioral intention has three dimensions including information disclosure, protection intention and transaction intention. The results show that privacy concern has a negative effect on information disclosure, but a positive effect on protection intention. The interactions of information sensitivity and compensation have significant effects on information disclosure and transaction intention. Additionally, information sensitivity has a negative effect on information disclosure and transaction intention. Therefore, separating a technology capable of inferring personal attributes from the societal implications inherent in that capability is essentially impossible. In the context of technology implementation and change management, the broader privacy and ethical issues become as important and difficult as the technical ones – if not more.

Consequently, as pointed out by Spangler et al. (2006), a detailed exploration of the ways in which data mining technologies can be used to collect, analyse and redistribute data is important not only because of the opportunities to enhance marketing/advertising efforts, but also because it sheds light on how consumers and society will react to the technologies – either positively or negatively. From a purely practical perspective, a negative reaction could cause consumers to turn away from the technology, the product and the company, thus counteracting any marketing improvements delivered by the technology. In that regard, knowledge of consumer and societal perceptions of privacy infringements is as important as knowledge of individual consumer demographics and buying habits. With that knowledge, companies can take measures to anticipate and prevent violations, and to compensate consumers in an appropriate manner when a violation of privacy is considered 'the cost of doing business'.

2.3.2.5 Media consumption

Despite all the research trying to understand what is the best advertisement to a particular user given a set of different contexts, it is the way it is presented that will influence him/her most. In this sense, the adopted multimedia delivery mechanisms play a preponderant role. Nevertheless, except for particular use cases, they are not specific to advertising. On the mobile sphere, Kim et al. (2009) presented a mobile advertisement system and a push scheduling scheme to deliver advertising message in systematic and effective way. The push advertisement scheduling scheme enables a system to select ad content guided by user preference models based on consistent characteristics subject to content category.

On the other hand, there are some research initiatives that study the best way to embed advertisements within multimedia files. The proposed systems, named ImageSense and VideoSense (Hua et al., 2008) are dedicated to images and videos by associating the most relevant ads with the most appropriate ad insertion points within the images or videos, respectively. The selection is not only based on relevance but also on visual similarity so that the ads yield contextual relevance to both the situation and the visual content. The ad insertion positions are detected based on visual saliency analysis to minimize the intrusiveness to the user. According to the authors, after passing through the contextual and inline types of advertising, this work allows entering the epoch of the third generation of multimedia advertising – impressionative advertising.

2.3.2.6 Summary

Altogether, as expected, each of the research initiatives only addresses part of the challenges, as they prefer to focus on specific topics around it. Furthermore, in the IT field, most works recur to advertising merely to use case the functionalities of the proposed solutions. In this sense, there is no end-to-end solution that allows a concrete study of this topic. However, each of these initiatives covers a particular issue relevant to the development and expansion of advertising industry, as they usually try to proportionate new advertising experiences. In addition, some works focus specifically on the necessary requirements to foster in this industry. Although we cannot say that the aforementioned research ideas failed to address the user and business partners' goals, due to their specificity, their contribution is limited. To get a better overview, Figure 2.14 depicts the most covered topics by the research and academia communities.

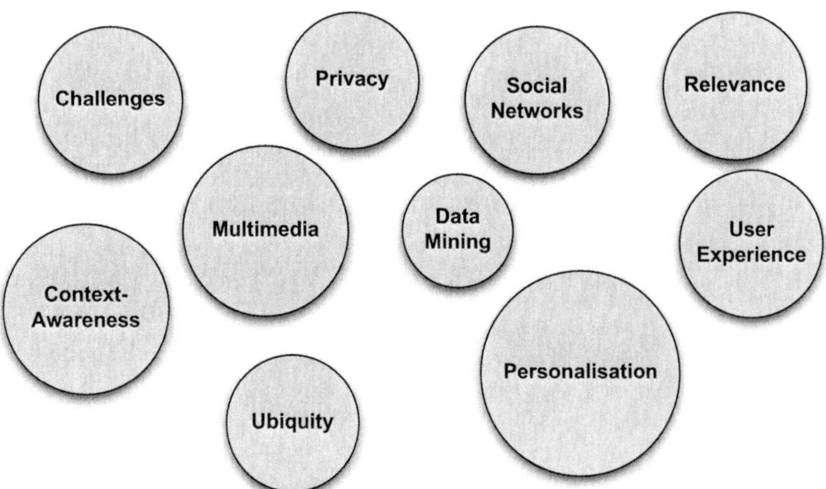

Figure 2.14 List of topics being addressed by research and academia

2.3.3 Industry

Fortunately, not only the academia but also some companies are looking towards this profitable market. In fact, they are the most active entities in this area. Unfortunately, due to the fierce competition in this field, most technical details of the commercially available solutions are usually omitted. Therefore, this section tries to cover the major capabilities provided by each product. This analysis will be divided into two types: the first covers specialized companies in providing advertising solutions within different multimedia domains, while the second deals with corporate solutions offered by major technology vendors.

2.3.3.1 Hosted advertising solutions

Within advertising scenarios, depending on the service, application or system being addressed, the adequate solution might require the provider to own an advertising platform or, in not so complex situations, leverage on existing ones. The following paragraphs deal with the second approach and are divided into remote advertising enablers (hosted and sharing resources) and advertising (ad) networks.

Representing the enablers, Acuity Mobile delivers the Embedded Mobile Advertising Platform (eMAP) together with Spot Relevance™ (Acuitymobile spot relevance http://www.mmaglobal.com/members/acuity-mobile-inc) and claims the ability to determine the right content, offer or advertisement, based on five dimensions: business goals, user activity, user interests, user actual and estimated location and time of the day. Based on a modular and scalable architecture, it allows seamless integration with existing marketing management systems and securely manages user data and identity information. Furthermore, besides supporting metrics collection for increasing the profiling relevance, it also allows both push (system initiated) and pull (user initiated) models, for advertising content delivery. On the other hand, Smaato provides the Open Mobile Advertising (SOMA) platform (www.smaato.com/soma), which alleges the delivery of targeted advertising to mobile phones within applications, on mobile websites/widgets and in mobile games (platform independent), connecting leading global mobile ad sales networks, with best-in-class mobile web and mobile application inventories, through their Software Development Kit (SDK). In addition, they support metrics collection that can be used to improve future campaigns. Another interesting solution is offered by iLoop (http://iloopmobile.com), which uses its mFinity™ mobile marketing platform to provide indirect delivery of targeted interactive text messages, containing links to a myriad of multimedia content. In addition, it features a complete Content Management System (CMS) that indexes, stores, manages and distributes mobile binary content including images, music, video, ringtones, wallpapers, themes, games, mobile applications and other digital content. Extra functionalities include reporting and an open SDK that allows third-party developers to integrate outside software into the platform.

In what concerns Ad networks, AdMob (http://www.admob.com/s/home) is the leading company. It offers optimized ad matching by storing and analysing data from every ad request, impression and click, providing advertisers with detailed metrics, which allow them to better understand the performance of their mobile advertising strategy. More concretely, for every ad request, AdMob analyses information available in the user's mobile browser (or embedded application). From this, AdMob determines device capabilities and more using open source tools and a variety of proprietary techniques. Another company worth mentioning is Quattro Wireless (www.quatrowireless.com), which besides offering standard ad metrics and mobile-specific data, through their dynamic targeting and ad serving Q-Elevation platform, ensures the right audience for each message, mapping behavior to discrete audience segments. It allows campaigns to be targeted based on the context values (e.g. media type, channels, publisher or site), demographics, mobile device, location and time. In addition, their ad programs and types include: application store optimization program, click-to-buy, rich-media, interstitial ad units, shake-up advertising, in-app peer-to-peer sharing, location awareness and user context database building.

Leading the world of modern advertising are companies like Google, Yahoo and Microsoft, which through their specialized advertising programs offer targeted display and search advertising for both web and mobile environments. Their biggest strength is to be able to index several types of information and therefore contextualize the results more accurately. Furthermore, they provide detailed metrics that help evaluating the success of the campaigns.

2.3.3.2 Dedicated advertising solutions

Mainly to support the needs of major telecom providers, dedicated products are sometimes a better alternative to a provide carrier-grade, flexible and scalable solution. Pioneering in this field, Ericsson offers two different products, the Ad Orchestrator and the Ad Broker (Asp et al., 2009). The first is a platform for buying, planning, delivering, monitoring and following up mobile advertising campaigns across multiple delivery channels. Further, it enables reach and relevance via a single unified platform and an easy-to-use graphical web interface. It is designed to work with existing infrastructures, enabling the monetization of companies own ad inventories, such as Short Messaging Service (SMS), Multimedia Messaging Service (MMS), mobile web, applications, ring-back tones, subscriber notifications and idle time, as well as to aggregate and monetize partner–publisher inventories. It features a modular, portable and layered architecture based on the JEE (Java Platform, Enterprise Edition) framework. The architecture has four logical layers, namely the presentation layer (i.e. graphical interfaces), core layer (business logic and associated components), integration layer (i.e. provide flexible integration mechanisms with associated components) and information layer (i.e. three logical data models: consumer, partner and campaign). Furthermore, it features a back-end system that allows data mining, billing, CRM integration, among other features. The Ad Broker is a further enhancement of Ericsson's advertisement solutions, making it easier for the advertisement industry to reach out, from one selling point, with campaigns that target consumers in multiple operator networks.

From Alcatel-Lucent, the approach resembles the previous one but focuses on the multi-screen experience (Fuerst & Fraser, 2008). Although this is not offered by a single product, the company vision goes towards that direction. On the mobile side, it partners with Placecast (http://www.1020.com/) and offers an end-to-end service that enables service providers to support brands in reaching their target audiences wherever they are, in innovative new ways (dedicated – virtualized – hosted solution). This carrier-grade solution (support millions of users) provides a unique ad engine to weave location information into highly relevant messages, enabling brands and advertising agencies with scalable, proximity marketing campaigns that will be pushed to consumers' mobile devices in SMS and MMS formats. The other solution is provided for the IPTV market. Based on the Alcatel-Lucent 7750 Service Router with Multi-service Integrated Service Adapter (MS-ISA, 8920 Audience Measurement Manager and 5930 Interactive Media Manager), it provides a robust advertising ecosystem for IP networks, with best-in class third parties, to support cue tone processing, campaign management, ad serving and ad operations (Alcatel-Lucent, 2010). By leveraging the transformed IP network and relationships with subscribers, it allows service providers to more efficiently deliver ads and help advertisers to reach the right consumers at the right time. With comprehensive tools for more precise measurement of ad performance and IPTV usage, service providers can offer their media partners insight into the consumer to better tailor their advertising and content offerings.

Another carrier-grade mobile advertising solution is provided by Comverse *(Get off to a quick start with Comverse mobile advertising!,* see References), which mainly targets network operators, whose offer includes triple and quad-play communication forms (IPTV, wireline, broadband, etc.). It is a comprehensive end-to-end in-network solution. Designed as a fully productized solution, it serves as the 'ad brain-engine' and uses the network for advanced targeting capabilities. As a modular, scalable and future-proof carrier-grade solution, it meets operators' growing needs for the long term. Furthermore, it allows the end-user to maintain control by using the operator self-service portal (which interacts with the ad brain-engine) and easily customize and manage their advertising preferences: how, when and what kind of advertising they would like to receive.

Also worth considering is the recently launched Oracle Communications Marketing and Advertising solution (http://www.oracle.com/us/industries/communications/046827.html), a scalable and comprehensive marketing and advertising management platform that enables network operators to enhance revenue and better monetize their networks. It delivers a comprehensive portfolio of interfaces, pre-defined templates and out-of-the-box applications that enable advertisers and agencies to create and schedule campaigns quickly and easily. It includes a Marketplace feature that matches advertiser and agency needs with inventory from publishers and the network operator, thereby facilitating the development of an ecosystem around which a network operator can build a sustained and thriving advertising offering. Furthermore, at the core, it possesses a powerful Ad engine which is integrated with the subscriber profile, notification and workflow, bulk messaging, traffic interceptor, monitoring, billing, and data abstraction services. These capabilities enable the delivery of targeted, personalized advertising and marketing campaigns utilizing multiple messaging channels, including SMS, MMS and Wireless Application Protocol (WAP) Push.

2.3.3.3 Summary

As it would be expectable, from the industry, we see a money-oriented trend. In other words, most work focuses on the mobile area and subsequent integration with existing advertising platforms, namely the web and TV. These tend to offer end-to-end solutions to accomplish the basic requirements for future advertising (see Chapter 3). These products are then divided into dedicated and hosted solutions. Among the first, we find mainly big manufacturers, while the second approach is usually dominated by mid-sized companies and different ad networks. Despite the differences between the different products, there are some common characteristics. In fact, most of them give particular emphasis to interoperability issues and the capability of integration and adaptation to multiple scenarios. Furthermore, they seem to have a similar goal: address the user needs and desires (i.e. in the simplest of the ways, by accounting users context information). Therefore, the ones that fail to understand these requirements will either be forced to act in partnership or work in a smaller scale (i.e. specific advertising solutions). Unfortunately, due to the fierce competition in this area, the technologies behind these products are usually hidden and the only information publicly available relates to the set of capabilities they support. Figure 2.15 tries to summarize some of the most addressed features.

2.4 SWOT analysis

Obviously not covering all the existing advertising standards, research or solutions available, section 2.3 gave an overview of the most relevant at the time of writing. Nevertheless, in order to assess the actual contribution/capabilities of each one of them, it is important to cross-check them with the advertising requirements

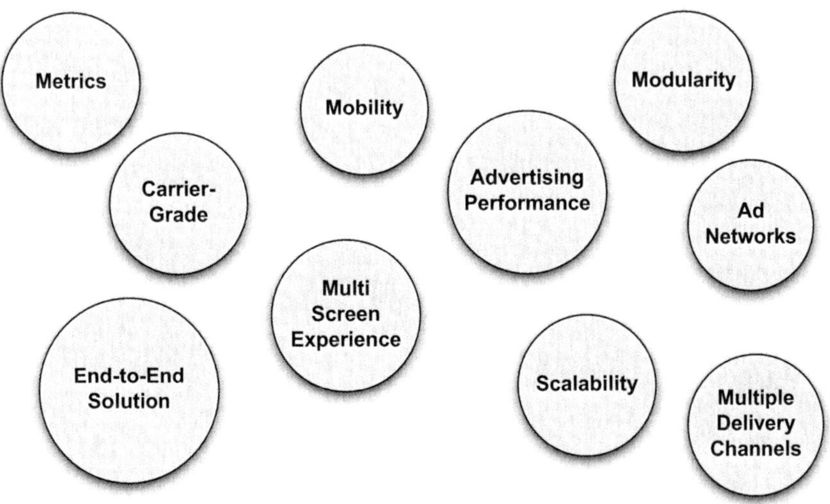

Figure 2.15 List of topics being addressed by the industry

identified in Chapter 3. This process will be done by doing a simplified Strength, Weakness, Opportunity and Threat (SWOT) analysis. In this sense, subsection 2.4.1 identifies, classifies and describes the works being compared. Furthermore, it presents a summary of the features/requirements that are supported by each of them. Then, after identifying the opportunities resulting from non-addressed requirements, subsection 2.4.2 describes how this work will work around the missing parts and finally, subsection 2.4.3 will cover some of the threats that might outcome from it.

2.4.1 Advertising approaches overview

To facilitate the comprehension when comparing the different approaches, the ad network approaches are grouped together because they represent a similar idea or provide identical features. Based on this assumption, Table 2.1 resumes the

Table 2.1 List of works to be compared with the requirements from chapter 3

Id	Categories	Description	Links and references
AdNet	Ad networks	Solutions offered by AdMob and Quattro	www.quatrrowireless.com/ http://www.admob.com/s/home/
MobAd	Enabler (standard)	OMA mobile advertising enabler	www.openmobilealliance.org
iLoop	Enabler (Product)	iLoop mFinity™ mobile marketing platform	http://iloopmobile.com
Acuity	Enabler (product)	Embedded mobile advertising platform (eMAP) + Spot Relevance™	http://www.mmaglobal.com/members/acuity-mobile-inc
Smaato	Enabler (product)	Open mobile advertising platform (SOMA)	http://www.smaato.com/soma
Comverse	Enabler (product)	Comverse mobile advertising	http://www.comverse.com/demo/mobileadvertising/
Ericsson	Enabler (product)	Ericsson advertising broker	(Asp et al., 2009)
Alcatel-Lucent	Enabler (product)	Alcatel-Lucent personalized and interactive IPTV advertising Solution + multi-screen experience	(Alcatel-Lucent, 2010; Fuerst & Fraser, 2008)
Oracle	Enabler (product)	Oracle communications marketing and advertising	http://www.oracle.com/us/industries/communications/046827.html
CUCAF	Enabler (prototype)	Key enablers for user-centric advertising across next generation networks	This book

approaches to be considered. Despite the importance of the research and other standardization initiatives, they were not included in this analysis because they did not represent an end-to-end advertising system.

In order to understand the actual functionalities of each solution, it is important to list a set of basic capabilities and requirements an advertising solution should possess (a detailed explanation regarding the requirements is given in Chapter 3). Once again, to simplify the visualization of the data, each of the capabilities/requirements will be assigned with an identifier to be used for the comparison. Table 2.2 contains this matching.

To provide an overview of the advantages and limitations of each solution, Table 2.3 introduces a 'big picture' comparison.

Table 2.2 List of expected capabilities/requirements and assigned identifiers, for an advertising solution

A. Cross-converged: web, TV and mobile advertising	
1. Channels	Does it support web, TV and mobile advertising?
2. Signalling	What is/are the signalling technologies used?
3. Delivery	Which delivery mechanisms are supported? (e.g. push and pull)
4. Transport	What are the multimedia delivery protocols it uses? (e.g. SMS, multicast,...)
5. Mobility	Does it support mobility? Which types? (e.g. session vs. terminal mobility)
6. Transcoding	Does it support transcoding? If yes, which formats? In real-time?
7. Splitting	Can you separate different media types from a single stream?
8. Adaptation	Can it adapt the media type according to the user device? What about other context information? (e.g. mute audio if environment is noisy)
B. Multi-context advertising experience	
1. Context parameters	Can campaigns be based on parameters other than demographics? Which parameters?
2. Reasoned parameters	Does it support second-level (reasoned) context as parameters?
3. Social	Can the platform infer social relationships and their strengths?
4. Predict	Can the platform suggest campaigns based on predictions of what users want, need or desire?
C. User interactive and controllable advertising experience	
1. Select device	Can users specify on which devices they want to be reached?
2. Type ads	Can users specify the type of ads they are willing to receive?
3. Profiling	Can users control the way they are being profiled? (e.g. on, off, weights, full)
4. Transparency	Are users aware of the profiling algorithms?
5. Policies	Can users specify what kind of context is to be disclosed to each of the different entities?
D. Flexibility to support different business models	
1. Players	Can different players in the advertising value of chain use the platform?
2. Granularity	Can you achieve a one-to-one marketing efficiency? (i.e. based on a set of advertisements existing at a given point in time, the platform should select which one is more appropriate to each user)

Table 2.3 Advantages vs. drawbacks of the advertising solutions being compared

Id	Advantages	Drawbacks
AdNet	Easy integration with applications and mobile websites, scalability	Limited cross-channel convergence, reduced context-aware experience, low user interactivity and control capabilities, linear business models
MobAd	Simple to use, supports video, scalability, easy integration with applications and mobile websites	Only mobile, no reasoning, limited context
iLoop	Flexible and scalable SMS marketing capabilities	Limited cross-channel convergence, reduced context-aware experience, low user interactivity and control capabilities, linear business models
Acuity	Good integration with mobile technologies, supports a reasonable amount of context information, reasoning, addresses different players in the value chain	Limited cross-channel convergence, low user interactivity and control capabilities
Smaato	Good integration with mobile technologies, supports several types of context information, addresses most players needs in the value chain	Limited cross-channel convergence, low user control capabilities, no reasoning
Comverse	Good cross-convergence capabilities, supports a reasonable amount of context information, existing user data exposure control, allows several business models	No reasoning
Ericsson	Excellent cross-convergence capabilities, supports a innumerable types of context information, reasoning, extensive support for business models	Limited user data exposure control
Alcatel-Lucent	Seamless multi-screen experience, fair support for different business models	Reduced context-aware experience, low user interactivity and control capabilities, no reasoning
Oracle	Good integration with mobile technologies, supports a reasonable amount of context information, addresses different players in the value chain	Limited cross-channel convergence, low user interactivity and control capabilities, no reasoning
CUCAS	Enables all the aforementioned advantages	Only a prototype and not tested commercially

Table 2.4 Overview of the main advertising solutions and supported functionalities

Requirements	AdNet	MobAd	iLoop	Acuity	Smaato
A.1 Channels	M-Web, App	Mobile, M-Web, App	Mobile, M-Web	Mobile, M-Web, App	Mobile, Web, App-
A.2 Signalling	HTTP	HTTP	HTTP	HTTP	HTTP
A.3 Delivery	Pull: Web	Pull: Web, App	Push: SMS	Pull: SMS, Web Push: SMS	Pull: SMS, Web, App Push: SMS
A.4 Transport	No media	HTTP Download	No media	No media	No media
A.5 Mobility	N/A	N/A	N/A	N/A	N/A
A.6 Transcoding	Banner Size	Yes	N/A	N/A	Banner Size
A.7 Splitting	N/A	N/A	N/A	N/A	N/A
A.8 Adaptation	Yes. Limited to device and Ad display	Yes. Limited to device and Ad display	N/A	N/A	Yes. Limited to device and Ad display
B.1 Context parameters	Media type, channels, publisher, site, device, geography, time of day	Device, geography, application	None	User interests, activity, geography, time of day	Carrier, content categories, device, profiles, geography, keywords, time of day
B.2 Reasoned parameters	None	None	None	Geography	None
B.3 Prediction	No	No	No	No	No
B.4 Social	No	Integrate with Social Networks	No	No	No
C.1 Select Device	No	No	No	No	No
C.2 Type ads	No	No	No	No	No
C.3 Profiling	No	No	No	No	No
C.4 Transparency	No	No	No	No	No
C.5 Policies	No	No	No	No	No
D.1 Players	Publishers, media agencies, application developers	Publishers, advertisers, media agencies,	Media agencies	Publishers, ad networks, application developers	Publishers, ad networks, application developers

(*Continues*)

Table 2.4 (Continued)

Requirements	AdNet	MobAd	iLoop	Acuity	Smaato
		application developers			
D.2 Granularity	Limited	Limited	Very Limited	Limited	Limited
Requirements	**Comverse**	**Ericsson**	**Alcatel-Lucent**	**Oracle**	**CUCAS**
A.1 Channels	Mobile, application	IPTV, mobile, M-web, application	IPTV, mobile	Mobile	IPTV, mobile, web
A.2 Signalling	Proprietary network delivery engine	HTTP, SIP	N/A	HTTP	HTTP, SIP
A.3 Delivery	Pull: SMS, ringback tones, interstitial ads Push: SMS	Pull: SMS, MMS, M-web, games, IPTV, M-TV, web, search engine, mobile broadband, App Push: SMS, MMS, e-mail, WAP	Pull: Video intersection Push: SMS, MMS	Pull: not specified Push: SMS, MMS, WAP	Pull: Push: SMS, MMS, E-mail, XMPP, video, Audio Push: SMS, MMS, e-mail, XMPP, Video, Audio
A.4 Transport	N/A	Unicast, multicast, cell broadcast	N/A	N/A	Unicast, multicast, overlay
A.5 Mobility	N/A	N/A	N/A	N/A	Session, terminal
A.6 Transcoding	Banner Size	No	No	N/A	No
A.7 Splitting	No	No	No	N/A	Yes
A.8 Adaptation	Yes. Not clear how	N/A	Yes. Not clear how	N/A	Yes
B.1 Context parameters	Channels, geography, browsing activity	Inventory context, usage patterns, consumer segments, device,	Geography, time of day	Geography, usage patterns	Usage patterns, geography, time of day, social networks, weather, traffic, device, consumer segments, dynamically defined parameters

		geography, time of day, social networks, dynamically defined parameters			
B.2 Reasoned parameters	None	Geography, billing data, etc. Connect with data mining software Ericsson CIM	None	None	Similarity, Influence, Matching, and other metrics provided by WeKa
B.3 Prediction	No	Situations, relationships	No	No	Situations, relationships, etc.
B.4 Social	No	Yes	No	No	Yes
Requirements	**Comverse**	**Ericsson**	**Alcatel-Lucent**	**Oracle**	**CUCAS**
C.1 Select device	No	No	No	No	Yes
C.2 Type ads	Yes	Limited	No	No	Yes
C.3 Profiling	On/Off, What, When	On/Off	On/Off	No	On/Off
C.4 Transparency	Limited	Limited	No	No	Yes
C.5 Policies	Limited	No	No	No	Yes
D.1 Players	Publishers, media agencies, operators	Publishers, ad networks, media agencies, application developers, network operators	Publishers, media agencies	Publishers, media agencies, network operators	Publishers, ad networks, media agencies, application developers, network operators
D.2 Granularity	Ok	Very high	Limited	Very limited	Very high

Putting everything together, Table 2.4 provides an overview of the main advertising solutions, concerning the type of functionalities they provide. For this analysis it is important to notice that the information was reviewed in November 2010. It may happen that in the meantime some of these solutions have been updated or discontinued. Furthermore, all the data that was collected is coming either from publicly available merchandizing or directly from a representative from the product/company.

2.4.1.1 Summary

Altogether, it is possible to understand that none of the solutions available is capable of providing all of the functionalities identified. In this sense, based on the analyses of Tables 2.3 and 2.4, the following topics need further improvement:

- *Convergence*: from the aforementioned solutions, only a few are actually trying to address a converged multimedia advertising experience. Nevertheless, most users still use different devices throughout an entire day. Therefore, the ability to provide a cross-device and network experience, with adapted multimedia content, will differentiate any service from the remaining competition. Furthermore, it will increase the reach of a campaign, which is usually a constraint for the advertising industry.
- *Context-awareness*: most enablers do provide some sort of user targeting based on different types of context. Nevertheless, some still focus on the old demographics approach, which is very limited. Trying to give some steps forward, only Ericsson is implementing some sort of prediction model that can help understanding situations and relationships. Therefore, it is expected that other approaches will focus on social context in a near future.
- *User experience*: there are many factors that influence user-perceived QoE. The most relevant is probably linked with the content itself, which is not under the control of the advertising enablers. However, there are other issues, which relate with the users' needs, willingness or desires. Nevertheless, few take this into consideration and do not provide any possibility for users to select or configure their personal preferences or privacy settings, respectively. Furthermore, most do not provide a transparent solution, where users can be aware of what information is being used by which entities. In a way, the trend will evolve from understanding customers to predicting what they want or need, but everything under a user-controlled environment.
- *Business models*: one of the major concerns when providing an advertising solution is to find the right target. Consequently, this is quite obvious for most approaches. Nevertheless, due to their architectural prepositions, some of them have a very restrictive actuation field. This means that some solutions, with exception from the carrier-grade ones, only target a small percentage of the value chain. This does not mean that they are not important or valid, but, they are limited to a smaller set of business models. On the other side, those who are able to target different players will have a great advantage to innovate

business models by leveraging new concepts to the different layers of the advertising value chain. Furthermore, currently, advertising is targeted according to what advertisers or media agencies want or plan. In the future, it should be perceived as a recommendation, that is, targeted to something users would want or need (i.e. according to a set of variables/contexts).

2.4.2 Strategy adopted

When looking at the available solutions in the market to address next generation advertising, it is possible to find several converging points. Nevertheless, most of the times, these only represent a specific set of functionalities. Furthermore, as seen before, none of the solutions identified seems to tackle all of the requirements defined for this work.

Therefore, this lack of systems integration opens an opportunity to explore the synergies when different technologies are applied together. Moreover, it is clear from the analysis that no product offers a reasonable way for users to control their own data and only a few allow the reasoning of information in real time to provide enhanced experiences. Unfortunately, the same applies with a cross-converged multimedia delivery solutions. In this sense, this work will focus on the interfaces between these three areas, which, as shown previously in this chapter, are the ones being researched, standardized and commercialized. The main innovations or differentiated factors should:

- Fill the gap between user personalization vs. privacy.
- Unify user-related data (context) in a single logical interface.
- Adopt user-centric approaches.
- Understand more about the users (e.g. see relationships).
- Infer new knowledge (e.g. predict current or future occurrences).
- Provide customizable products addressing the needs of different players.
- Optimize the relationship of cost/efficiency.
- Promote innovation for the industry: new business models/new experiences.

and perhaps the most important:

> Apply all of the aforementioned principles into a converged system, capable of addressing the heterogeneity of the advertising industry in terms of: networks, devices, players, business models, specifications, systems and users.

A more detailed overview of the strategy adopted is given in Chapter 4, while a specific description and modelling of the specific enablers is given in Chapters 5–7. However, like any other approach, there are some limitations and threats that need to be taken in consideration. In this sense, the next subsection highlights some of these issues.

2.4.3 Threats and risk assessment

In a world ruled by enterprises more than governments, there is always a risk to compete against the industry giants. In this sense, this work's posture is slightly different. It can be seen as a complement and not as a threat. Still, due to the important role advertising plays in today's society, it is very likely that companies will invest a lot of money (and consequently human effort) in this field. Therefore, it is not expected that this work will outperform some of the existing solutions. It will, on the other hand, focus on the integration of different technologies to address the issues not yet tackled by the industry.

In addition, due to the lobby that might exist in the industry, there is the possibility that each player will want that business models remain as they are. In this sense, even though current or future technology can be used to improve the overall value chain, some aspects might be ignored. At the same time, the lobby can influence regulators to limit the 'expansion plans' for the new players. This is particularly important for the entity managing the user-related contexts, which raises notable privacy and security considerations.

More important is the fact small players will never be able to roll out such features, as they do not possess enough reach. In this sense, an entire framework covering the previously mentioned opportunities is likely to be feasible for a few players. Nevertheless, smaller components can be developed and collaborative partnerships can be achieved.

Another concern relates with the execution of this work, as it is possible that in the meantime, the industry changes in different directions, making this work unusable or not applicable. In this sense, to avoid obsolescence, this work tries to follow standards and recommendations whenever possible. When not, it will try to integrate things in this direction. This will reduce the risk of incompatibility with future technological advances.

Chapter 3

Requirement analysis

After highlighting the motivation of this work and over viewing the advertising state-of-the-art technologies, it is important to focus on the aspects that will provide the guidelines for this work. In this sense, this chapter presents the user requirements, which deal with the users' expectations and business partners' requirements, covering the main issues that might influence the technology adoption and the functional requirements, which relate to the system functionalities and the non-functional requirements, dealing with other important aspects for the proposed advertising framework. Figure 3.1 illustrates the relationships between the different requirements.

3.1 User requirements

Assuming from the beginning that the main motivation relates to the improvement of the user-perceived QoE within advertising scenarios, it is extremely important to understand what are the user requirements to provide an optimal advertising experience.

Therefore, besides collecting some of the results provided by other research initiatives (Eriksson & Akesson, 2008; Ranganathan & Campbell, 2002), under the scope of this book, a survey to evaluate what were the characteristics users would value the most in a future advertising system was developed. The findings were documented in Simoes and Magedanz (2009). When compared with other works, it was possible to find interesting correlations among the most valuable aspects. Table 3.1 lists the most common user requirements.

It is important to notice that some of them were not specifically identified for advertising, but for next generation services in general. Nevertheless, it makes sense to adapt these to advertising as this work foresees it fostering as a service itself, in a near future.

The following section presents a similar overview but from the business partners' point of view. Obviously, with the purpose of targeting users' expectations in the best way, there are other requirements that may influence the decision of whether or not these players will enter the new era of advertising.

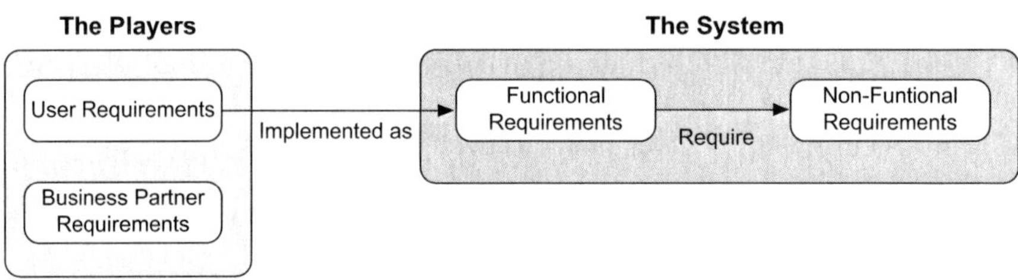

Figure 3.1 Relationship between the different requirements

Table 3.1 List of most desirable requirements for future advertising scenarios

Id	Requirement	Description
R.1	Personalization	Ability to deliver personalized media (services or contents) to the final user by matching their preferences and interests into a global context, providing at the end, a final service tailored to address the specific needs, desires or intents of the consumer.
R.2	Contextualization	Capability to match contents and services to any user-related context (personal, social, temporal, environmental, etc.).
R.3	Adaptation	Possibility to adapt services and contents according to the type of user, environment and different end terminal used to consume the media.
R.4	Interactivity	Enhancing the way a service is consumed. Allowing the user to interact with a service creates a whole new world of possibilities and business scenarios.
R.5	Mobility	Describes one's ability to use technology while moving. This will enable ubiquitous and pervasive computing scenarios where users experience services anywhere, anyhow and at any time.
R.6	Privacy	Capability for the user to possess and control their own data. Moreover, it should have the right to dispose and distribute it when he finds necessary. Possibility to opt-in and opt-out.
R.7	Relevancy	Ability to deliver something that fits the appropriate context and meets the user expectations, needs or desires.
R.8	Transparency	The act of making clear to the user how the system/service works and which data is being used or exposed.
R.9	Rewarding	Independently from the way the service is presented, it brings some value to the user. It can happen in the form of an offer, discount, etc.

3.2 Business partners' requirements

Despite this book focusing mainly on the user side of the advertising world, it would be naive if the system would not consider business-related requirements. These involve several stakeholders, from manufacturers, network operators, passing through service providers, until advertising and marketing agencies (see section 2.1.5 for more detail). Although these requirements may not fit all the needs of each entity, they represent a good overview of the factors that may lead to success in next generation ubiquitous and heterogeneous advertising environments (Mobile Codes Consortium, 2008).

B.1 – Innovative business models

Companies that want to take advantage of the opportunities presented by new technology will have to understand how it may fit with existing business models or how it can create new ones. Mobile phone manufacturers and application developers have to consider that they have the opportunity to become providers of services (e.g. as context providers), as well as providers of products. These should not only suit the needs of the market, but also drive the market in new directions, with open architectures and tools, allowing new applications to develop, based on their devices and platforms. Internet providers and mobile network operators need to consider how advertising can be integrated with existing tools for revenue generation such as SMS, Premium SMS and MMS, as well as the huge potential to drive data traffic (which in most markets is still priced highly and discourages usage increase). Additionally, the fact that operators have an existing billing structure is of great advantage, particularly, as far as payments for low amounts are concerned. Finally, different entities need to understand how to collaborate with different players, bearing in mind that in future advertising applications, they will only be one part of the solution.

B.2 – Consumer adoption

Independently from the service of product an industry is promoting, its success is always dependent on a common factor, its customers. In this sense, the future of advertising is not an exception. Therefore, the possibility to experiment it should be available to everyone, anywhere and anytime. Nevertheless, for that to happen, they will need to be able to use these services with any device they buy, whatever kind of subscription or contract they have signed, independently from operator they have chosen. For advertisers and remaining players in the value chain, this means that choices need to be made to support interoperability and openness, rather than to develop and use private or proprietary models. Once there is a global, interoperable platform available, the different players will be able to focus their resources on creating and launching new services or on transferring existing services to distinct devices, all based on a system that is trusted by the consumers.

B.3 – Technology availability

Without certain technologies, ubiquitous advertising over heterogeneous environments is impossible. The full potential of a mobile device, as an ubiquitous object, is only released when local interaction technologies are embedded and combined with external systems. The integration of secure mechanisms will be of great value to provide security and trust for handling valuable data to the outside world. Although not addressed by this work, the support for high-quality display of the information to the users, supporting state-of-the-art Internet technologies, is required to provide satisfying and consistent user experiences. Likewise, mobile networks need to be able to deal with large amount of traffic at high speed. Therefore, networks should be optimized for multi-group multimedia delivery. To sum up, the availability of technologies will be vital to the success of personalized multimedia advertising.

B.4 – Interoperable systems

Interoperable systems are essential to mass adoption of an advertising platform. Though proprietary systems may function well in limited situations, they are a barrier to any large-scale implementation. Therefore, standards should be used when possible to avoid compatibility issues. Furthermore, the data models used to represent advertisements should be open, flexible and inter-changeable. This will allow standardized exchange of information between different entities in the value chain of advertising.

The next sections show how users and business partners' needs can be materialized into functional and non-functional requirements. Furthermore, it divides them into three different categories, which are associated with the adopted work methodology, to improve the way users will perceive advertising in the future.

3.3 Functional requirements

The functional requirements describe the capabilities that the proposed framework should implement. In this section, they will be presented and classified according to three distinct but complementary categories.

3.3.1 The human side of information technology systems

A.1 – Flexible user context and data management system

The system needs to support different types of user-related context. These should include: user profile (e.g. birthday, preferences), social networks (e.g. list of friends, affiliations), devices (e.g. brand, resolution), external/environmental context (e.g. weather, location), reasoned knowledge (e.g. mood, recommendations), profiling (e.g. on, off, manual) and policies (e.g. who, when, how) among others.

A.2 – Sophisticated user context and data exposure layer

The previously described information must have a way to be exposed to third party applications. It should not only be possible to request information on demand, but also subscribe for notifications, whenever a context update occurs. In both cases, it should be allowed to perform queries based on users or context types, ideally using regular expressions.

A.3 – Enhanced security, privacy and trust mechanisms

Due to the sensitivity of the data being managed by such solution, it is mandatory that the entities requesting user-related context data are duly authenticated and authorized. Moreover, it should be clear in any situation how the data can be handled by third party providers.

3.3.2 Understanding and predicting human behaviour

B.1 – Generic user data modelling framework

This framework should comprise a series of steps that should be performed in order to finally model human behaviour. These should deal with original data problems (pre-processing, which usually involves data cleaning and preparation) and provide a generic way to test different data mining algorithms (common or self-made ones).

B.2 – Advanced user data reasoning framework

Similarly to the modelling framework, this should offer a way to combine different types of context information, find correlations among them and infer new data based on past validated experiences.

B.3 – Classification and quantification of user-related inferred data

To allow the reuse of user-related inferred data, it is necessary to specify a way that will enable third parties to use this information. In this sense, there must be a syntax that allows a generic definition for the reasoned data quality, or a degree of certainty for such prediction.

3.3.3 Context-aware multimedia management system

C.1 – Personalized multimedia management towards users

It should be possible to start, modify and terminate multimedia sessions towards different users in a personalized and adapted way (according to his/her contexts and devices), independently from the access technology being used. The system should automatically detect the best conditions to perform each of the aforementioned tasks. Moreover, it should support mobility scenarios (handovers), without significantly impacting the user QoE.

C.2 – Efficient multimedia management towards content providers

Depending on the capabilities of existing content providers and streaming solutions, the system should be capable of requesting multimedia sessions to be initiated, modified and terminated.

C.3 – Flexible context-aware delivery system

It should be possible to automatically react to context changes. In this sense, the system must be able to act based on a rule-based system, enabling at the end, context-based, real-time content delivery and adaptation.

C.4 – Provide adaptation at session, network and transport layers

Using the previously identified mechanisms, the system should provide interfaces to allow context information to be passed to external entities responsible for performing operations at the session, network and transport layers. These may include resource reservation, Quality of Service (QoS) enforcement and network selection among others. In other words, it should provide mechanisms to improve the overall efficiency of the multimedia delivery solution.

3.3.4 Advertising framework

D.1 – Cross-converged: web, TV and mobile advertising

An advertising platform that is capable of seamlessly addressing cross-converged environments, namely web, TV and mobile, should be created by adapting multimedia advertising content accordingly. Furthermore, it should support mobility scenarios between different access networks for each of the aforementioned solutions.

D.2 – Multi-context advertising campaigns

It should be possible to assign advanced marketing campaigns, which are not exclusively based on demographics, but instead, on real-time data (context). In other words, it should be possible to choose from a list of context providers, which type of data the advertisement should react to and in which way. Obviously, these must be in conformance with both user and operator (when applicable) self-defined policies.

D.3 – User interactive and controllable advertising experience

Users should specify in which way they can be reached. They should be able to choose what type of information can be used and that only some people (advertisers) can contact them. Furthermore, they need a way to specify which type of ads they are interested (discounts, offers, propaganda) and even influence the way they receive it (e.g. by specifying preferred devices based on context information). Ideally, they should also be given the opportunity to control the way they are being profiled (change the algorithms provided by the ad agency or at least modify some of its parameters).

D.4 – Serendipitous advertising

Advertising should become spontaneous, creating a ubiquitous and pervasive platform that takes user requirements under consideration. It should enable the triggering of advertising based on user context. It should be perceived more as a recommendation.

D.5 – Addressing users needs, desires or intentions

Based on reasoned data (with or without certainties), the platform should be able to suggest interesting advertising to the users. This should interact with the campaign management interface, allowing advertisers to see what users are looking for. Basically, this intelligent system should represent a win-win model to both advertisers and users.

D.6 – Flexibility to support different business scenarios

In an advanced advertising solution, it should be possible that different business actors provide different components of the multi-context enhanced advertising functionalities. Therefore, the architecture needs to enable cooperation between different business entities and should be based on open interfaces. Furthermore, it should allow targeting both, broad audiences with similar interests and a single user with specific requirements (everybody does not like the same things, and even if they do, they might not want them at the same time).

3.4 Non-functional requirements

All system components that support and enable the integration or interworking of different advertising-related technologies need to be integrated into a common multi-context advertising system architecture. Such an architecture needs to fulfil several requirements.

3.4.1 Usability

Usability is a measure of how well the software supports the execution of user tasks. Key factors contributing to usability are the presentation of information and the management of user interaction (in this case it also includes the advertiser interface). Usability includes two other measures, affordance and accessibility (Greenfield & Short, 2004). *Affordance* is a measure of the cost of learning to be productive with a user interface. A user interface may be highly usable once it has been learned, but difficult to learn. A user interface that is highly intuitive, and therefore easy to learn, is said to have high affordance. *Accessibility* is a measure of how broad a pool of users can interact effectively with the user interface. A highly accessible user interface can accommodate the needs of users with a wide variety of requirements.

3.4.2 Performance

Performance is a measure of how quickly the system responds to stimuli and how well it utilizes resources in providing that response. Performance is often defined in terms of latency, throughput, efficiency and scalability (Grady, 1992). *Latency* is a measure of the amount of time that elapses in performing a given operation under a given operating load. *Throughput* is a measure of how many operations can be performed in a given amount of time under a given operating load. *Efficiency* is a measure of how many resources must be consumed by the software to provide acceptable latency and throughput under a given operating load. *Scalability* is a measure of how many additional resources must be consumed by the software to maintain acceptable latency and throughput with increasing load.

3.4.3 Reliability

Reliability is a measure of the frequency and severity of defects encountered during normal operation of the software. The more severe the defects and the more frequently they are encountered, the less reliable the software. Three related measures are: fault tolerance, robustness and security (Grady, 1992). *Fault tolerance* is a measure of how well the system can maintain normal operation when defects are encountered. Examples of fault tolerance in distributed business applications include recovering from failed transactions and routing tasks to other resources, when the original resources are incapacitated. *Robustness* is a measure of how well the system avoids failures when confronted with invalid data or incorrect usage. *Security* is a measure of how well the system avoids a specific type of failure, namely the unauthorized exposure of the processes and entities manipulated by the system.

3.4.4 Supportability

Supportability is a measure of the cost of supporting the software after it has been delivered to the customer. Measures that contribute to supportability include maintainability, malleability, extensibility, portability, interoperability and testability. Nevertheless, only some of them apply to the current situation (Greenfield & Short, 2004). *Malleability* is a measure of how easy it is to modify the software to accommodate changes in requirements. This is determined by the degree of encapsulation of the components, and by how well they can vary independently of one another. *Extensibility* is a measure of how easy it is to replace existing parts of the software, and how easy it is to add new functionalities. This is determined by the degree of coupling among the components. When components are highly coupled, it is difficult to replace one without affecting the others. *Portability* is a measure of how easy it is to adapt the software to run on different platforms. This is determined by how well platform dependencies are localized and encapsulated within the software. *Interoperability* is a measure of how easy it is to compose the software with other systems. This is determined by how well the software exposes its functionality through programmatic interfaces and how much context must be maintained by the other systems to use those interfaces.

Chapter 4

Designing a converged user-centric advertising framework (CUCAF)

Chapter 3 provided the basic requirements for this work, and consequently the guidelines for the system architecture. Together with the methodology introduced in section 1.4, this chapter covers the design principles of the *Converged User-Centric Advertising Framework* (CUCAF) and consequently its architecture. Furthermore, it shows how the same system (the implementation of the architecture) will be evaluated. Nevertheless, beforehand, it is important to specify the adopted model and respective functional entities.

4.1 System model and functional entities

This section defines the system architecture and functions for the CUCAF. By definition, it is a multi-advertising framework based on different communication technologies and context types. It, therefore, introduces generic communication and context abstraction layers, in order to provide a uniform platform, applicable to all different technologies. These abstractions are largely based on those that are specified by the Third Generation Partnership Project (3GPP), for example the IP Multimedia Subsystem (IMS) or the Evolved Packet Core (EPC), and the ones defined by some context-aware research initiatives, respectively. Therefore, whenever possible, the terminology used will remain the same. Since the overall scope of this work is covering only parts of the topics investigated in the related work (see Chapter 2), in here, the communication and context abstractions, as well as functions, have been simplified, when possible.

4.1.1 The communication layer

This project considers a communication layer that enables communication between two end-systems, which are denoted as the user end-system and its corresponding advertising end-system. These end-systems are located in a User Network (UN) and an Advertising Network (AN), respectively. Each end-system contains a service application function (Appl.). Between the service application functions of these end-systems, a communication session runs via a bearer. This bearer is provided by session functions at the Bearer End-Points (BEPs). A bearer is transported via flows. Flows are connectivity elements that are provided by a certain communication technology, where the network, transport and physical realization of a flow differs in different technologies. A flow requires three features:

- A *flow identifier* from a (technology specific) flow identifier namespace. Flow identifiers can be either a single identification object, or alternatively, they can be a set of locators[1] of the Flow End-Points (FEPs), like source and destination addresses in Internet Protocol (IP). The flow identifier can also contain a flow classifier, which characterizes the required treatment of the flow along the communication path and in the FEPs.
- A *routing mechanism* that is based on the flow identifier.
- A mechanism to associate *service requirements* with a flow.

A flow is a unidirectional communication element. Some technologies automatically establish bidirectional communication by means of a pair of flows. A flow is limited to the domain where a common flow identifier namespace is used. At the boundary between different namespace domains, flows are terminated and mapped to a corresponding flow in the other domain. By that, end-to-end connectivity (communication) for the bearer is provided by a series of flows, as shown in Figure 4.1. Boundaries of the flow identifier namespace typically correspond with technology boundaries. For example one end-system can be located in an IPv4 network, and the corresponding end-system is located in an IPv6 network. In this case, an IPv4 flow establishes the connectivity in the IPv4 domain, an IPv6 flow establishes the connectivity in the IPv6 domain, and at the domain boundary, these flows are mapped to each other (a similar situation could occur between unicast and multicast capable technologies).

A flow can span over network boundaries if they share the same flow identifier namespace. A flow can have multiple end-points, thus establishing a point-to-multipoint connectivity for multicast services. These bearers are bound to a flow by session, network and transport functions in the end-system, while the flow-setup

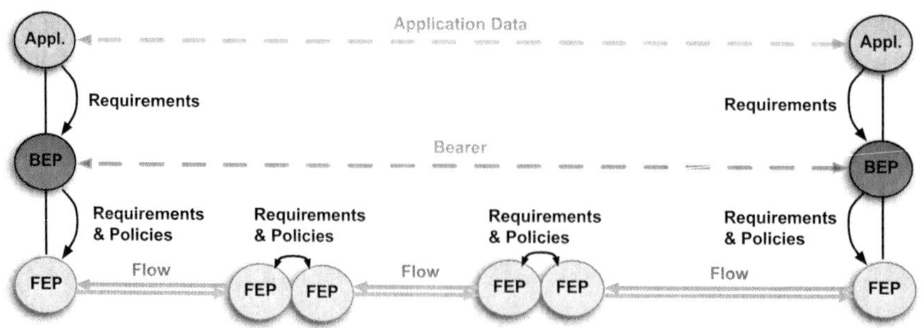

Figure 4.1 Relationship of bearer and flow connectivity elements with local bearer management

[1] A locator is the identifier of the location within the routing topology.

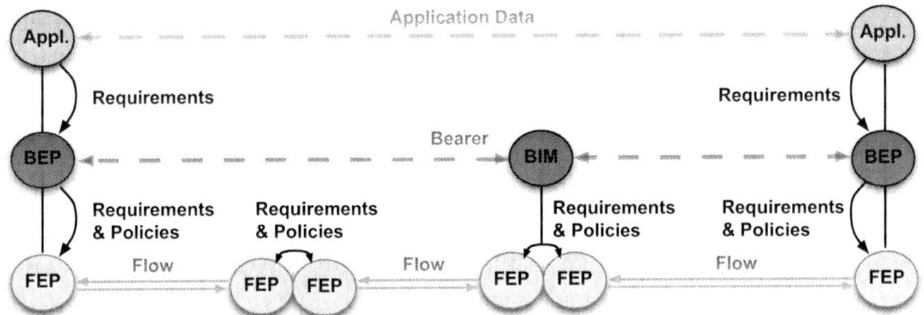

Figure 4.2 Relationship of bearer and flow connectivity elements with distributed bearer management

Application Programming Interface (API) provides some means for the configuration of the flow (e.g. for QoS parameters).

An alternative approach is distributed bearer management, as depicted in Figure 4.2. In this case, bearer management functions are located not only at the BEPs, but also within at least some of the intermediate networks. Bearer management signalling is used between the bearer management functions within the end-systems and Bearer Intermediaries (BIMs). Common bearer management signalling protocols are the SIP (Rosenberg et al., 2002) and the Service Delivery Platform (SDP) from the Internet engineering task force, which describe and negotiate the requirements and parameters of multimedia applications, such as audio and video. A bearer intermediary is then an SIP function located in the network, for example in the IMS or EPC. Despite the illustrations above which show the communication model used in this work, this book focuses mostly on the upper layers (nevertheless, this subject is further referenced in Chapter 7).

Using a simplified model, it is possible to identify three different types of functional entities, the *user*, the *operator* and the *advertiser*. The user is the individual who will benefit or consume the content owned by the advertiser. The operator is the entity that is responsible for enabling the communications (including signalling and media) between the user and the advertiser. The advertiser is the entity that pretends to distribute content to the users. However, the entity operator can be further divided into connectivity abstraction and session management entities. In this sense, the relationships between the functional entities for the communication model are represented in Figure 4.3.

The main reason why these separations exist relates to the multi-layered approach this project wants to impose inside the architecture. This increases flexibility and scalability, which are requirements for this framework. Furthermore, it allows the session layer to partially control the network and transport layers.

4.1.2 The context layer

Context, 'embedded' into our world and around our lives, can be exploited to enrich all types of communication systems (especially mobile). To enable such

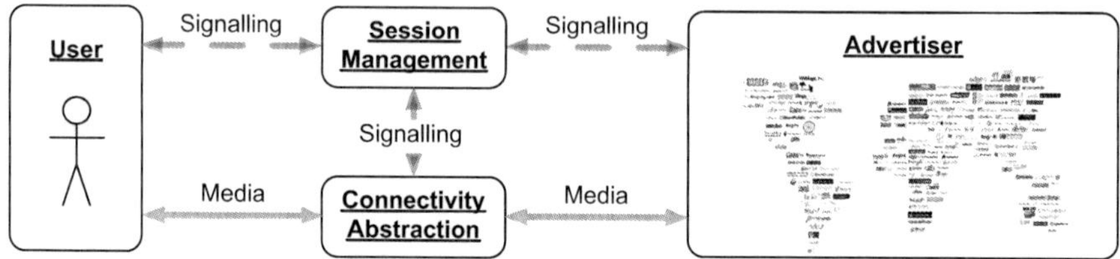

Figure 4.3 Relationship between the different functional entities regarding the connectivity layer

vision, it is necessary to have a model that considers context acquisition, management, consumption and distribution. A well-known and often used paradigm within networked-based services is the producer–consumer role model (Murphy, 2000). The proposed model is based on such paradigm and is built upon three basic functional entities: *context consumer (CxC), context broker (CxB)* and *context provider (CxP)*. Figure 4.4 depicts the relationships between these entities.

In this scheme, the context producers and consumers are decoupled in space and time; therefore producers can continuously (non-blocking) generate context, while consumers can consume the context on request or based on subscriptions. The decoupling is important as it impacts on the overall scalability of the system.

- *Context broker (CxB)*: the core component in this architecture is a centralized CxB. It works as a handler of context data and as an interface between other architectural components. The broker controls the flow of context information and maintains information about all active CxPs by means of an announcement process. In addition to offer a context-provider lookup service, the CxB

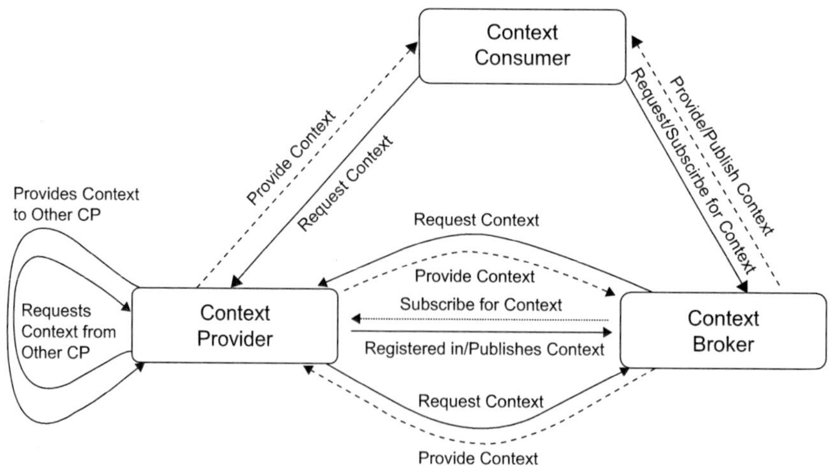

Figure 4.4 Relationship between the different functional entities present in a context-aware model

maintains context cache and provides access to old context information via a context history service.

- *Context provider (CxP)*: is a module that provides context information in synchronous mode; that means that a CxC and the CxB can invoke the CxP in order to acquire context information. A CxP can only provide context data specific to an invocation; it never sends data to another platform component in asynchronous mode. Moreover, a CxP can produce new or high-level context information, inferred from low-level context information or sensor data. Every CxP registers its availability and capabilities by sending appropriate announcements to the CxB and exposes interfaces to provide context information to the CxB and/or to the CxC.
- *Context consumer (CxC)*: is an entity (a context based application) that uses context data. It can retrieve context information, either by sending a request to the CxB and subscribing to a particular context event, or by invoking the CxP directly over a specific interface. The CxP interface can be requested from the CxB.

Just like in the connection layer scenario, both the user and the advertiser represent functional entities. The most common situation would be that the user behaves as a CxP, obviously not directly, but maybe through his/her device. Even when not representing a content provider itself, most of the times he/she is the source for a context provider (e.g. location context provider states where the user is). Consequently, the advertiser would be acting as a CxC, requesting user-related context to adapt or personalize the advertisement accordingly.

4.1.2.1 Context modelling and representation

The context information needs to be represented and modelled for being machine interpretable and exchangeable using well-defined interfaces. The goals are to support easy manipulation on many devices, including resource constrained mobile devices, achieve low overhead in keeping the model up-to-date, easy extension, efficient search and query access, having cheap and simple mechanisms for adding new types of information and scalability. In the literature, different approaches for representing contextual knowledge and semantic information can be found (Baldauf et al., 2007). A richer representation model enables complex and intuitive reasoning. Moreover, due to the assumption that all user-related information will be dealt as context (e.g. preferences, device, social, environment and policies information), this representation becomes even more critical.

In order to provide a lightweight and agile context encoding and communication system, heavy ontologies might not be an appropriate choice for a very fast system, handling a lot of context data. A more appropriate option for representation and exchange of context information is the *ContextML* (see Knappmeyer et al., 2009, 2010a, 2010b and the C-CAST project http://www.icttcast.eu/for specification details) based on the XML format. It is used in the presented architecture to announce the type of context (e.g. language) and the provided scope (e.g. Portugal as a country or location). Furthermore it is used to represent time constraints such as context creation time and period of validity.

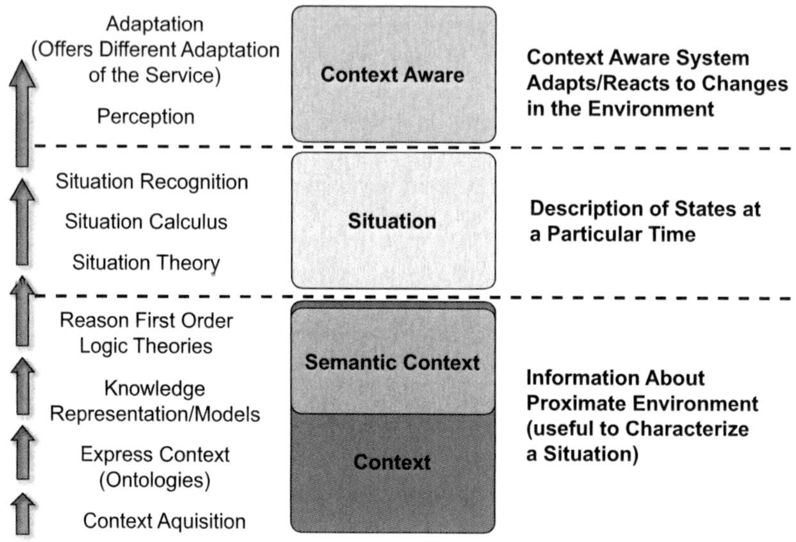

Figure 4.5 Context layering model [Adapted from Baker et al. (2009)]

As illustrated in Figure 4.5, there are several layers of abstraction in a context-aware system and any context-aware middleware, or architecture, must therefore be capable of building representations and models of these abstractions. However, these high-level abstractions can only be made from lower level context, which requires some form of context management function (performed by the CxB). The main context management features are context acquisition, context aggregation and fusion, context dissemination, discovery and lookup.

In order to manipulate context information, it must be represented in some form that is compatible with the models that will be used in the reasoning and situation recognition processes. These models could be object oriented, ontological, rule based, logic based and based on semantic graphs or fuzzy logic sets. Expressing context using just one representation is almost impossible since the range is from the most specific, for example a temperature reading, to the most abstract, the state of happiness. Furthermore, the representation must lend itself to the reasoning and inference techniques to be used, such as classification, filtering, aggregation, feature extraction, taxonomies, data mining, clustering, pattern recognition and prediction. Reasoning mechanisms allow high-level context to be deduced or situations to be recognized and often the output of one process can be used as an input to another. Moreover, reasoning is also used to check the consistency of context and context models.

4.2 CUCAF: converged user-centric advertising framework

Like in any other service, effective advertising needs to know what users want, need and desire to be able to target them in the best possible way (increase their

QoE). Figure 4.6 depicts the logical steps to achieve such goal. First, it is necessary to collect some information about them. In this sense, it is essential to find an efficient way to manage user-related heterogeneous data (coming from different sources). All this information is called *human context*. Once this is done, the information needs to be processed in order to better understand how users think, namely, what, why, when and how they do things.

This can be done by using reasoning or inference techniques, which can be used to group data, find patterns or predict future events, among others. Obviously, when specific models are applied (others than common statistical and data mining), new sorts of knowledge are also expectable. Knowing more about people allows advertisers to adapt their goals according to user contexts (both simple and reasoned), enabling a more personalized service experience. This process is denoted as *understand and predict*. Nevertheless, these data are only useful if used properly to interact with the users. Independently from the form this interaction takes place, it is necessary to deliver multimedia content (advertisements) in the best possible way (again, using user context). This process is extremely important as it immediately impacts the user-perceived experience. Due to its broadness, it was named as *multimedia delivery*.

Altogether and when combined, these three stages will represent the basic functions that the proposed framework should offer. Figure 4.7 highlights the relationship between these, the functional entities and the abstraction layers presented in section 4.1. The realization of such concepts into components will be introduced in section 4.3.

Figure 4.6 Complementary stages inherent to the advertising framework

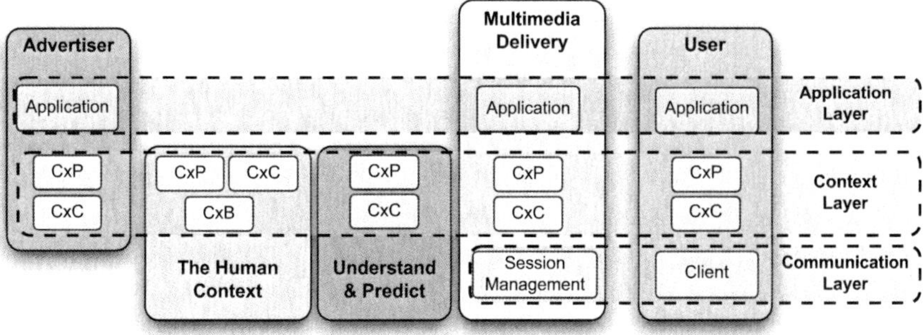

Figure 4.7 Relationship between logical processes, entities and abstraction layers

4.3 CUCAF architecture

4.3.1 Introduction

Chapter 3 defined the general requirements for future advertising solutions across heterogeneous environments (e.g. devices, networks, contexts). However, from the work collected in Chapter 2, it is clear that there is no system available that is capable of matching the aforementioned requirements. Therefore, in trying to overcome these issues, this book introduces a multi-context enhanced advertising framework called Converged User-Centric Advertising Framework (CUCAF), which represents the basis for this work.

The different steps for the management of advertising across heterogeneous environments are the human context, understanding and predicting human behaviour and efficient multimedia delivery (see Figure 4.8). Before investigating those contributions individually in later chapters, this section discusses how these functions can be embedded into a common multi-context enhanced multimedia delivery system architecture (advertising framework). Furthermore, based on the derived requirements and functional components, it presents and analyses different alternatives for mapping the functional components into a network/system architecture.

4.3.2 CUCAF functional reference architecture

The CUCAF architecture can be described based on a reference architecture, which specifies the functional entities. By separating the solution into three

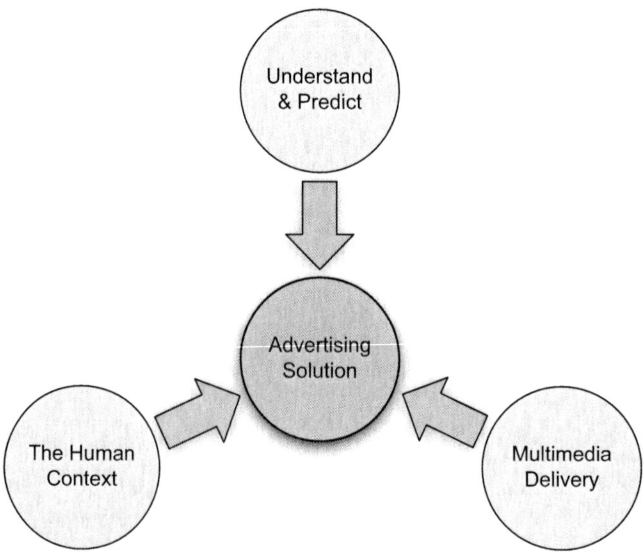

Figure 4.8 Combination of the different steps for managing advertising across heterogeneous environments

complementary enablers, this approach reduces deployment efforts by means of reusing common function components and allows interoperability across a wide variety of environments in a consistent manner. In fact, the implications of this decision go well beyond technical aspects. From a business perspective, it accelerates the deployment of new services by leveraging the interaction between components and applications developed by different providers. For this same reason, within this book, instead of developing new software, external calls to other enablers are performed.

In this sense, the functional reference architecture is depicted in Figure 4.9 and contains three main functional entities, which implement the steps depicted in Figure 4.8. The central data (context) management entity is the *human enabler*. Its capabilities are divided into four adjacent modules, which are responsible for controlling, managing and securing context information. The Human Context Broker (*HCB*) deals with all the requests coming from the outside (other components), while the *Human Data Repository (HDR)* is responsible for storing user-related context data. The Policy Evaluation, Enforcement and Management Architecture (*PEEM*) and *authentication* modules are responsible for assuring authorization and authentication, respectively. The *reasoning enabler* comprises an engine that coordinates a series of steps that allow data to be prepared, processed and evaluated in real time. These data usually try to understand or predict human behaviour and therefore, the outcome of these interactions results in new knowledge that is duly stored in the HDR (through the HCB). With the task of managing all advertisers to user interactions (and vice-versa), the *Session Management Enabler* (SME) is responsible for distributing the right content to the right users, according to specific context information. While the *signalling module* deals with all the Session Initiation Protocol (SIP) and SDP control messages used to manage sessions between the content and the user, the *Content Processor and Delivery (CtPD)* has the task to stream/provide the content to the user, using the RTP

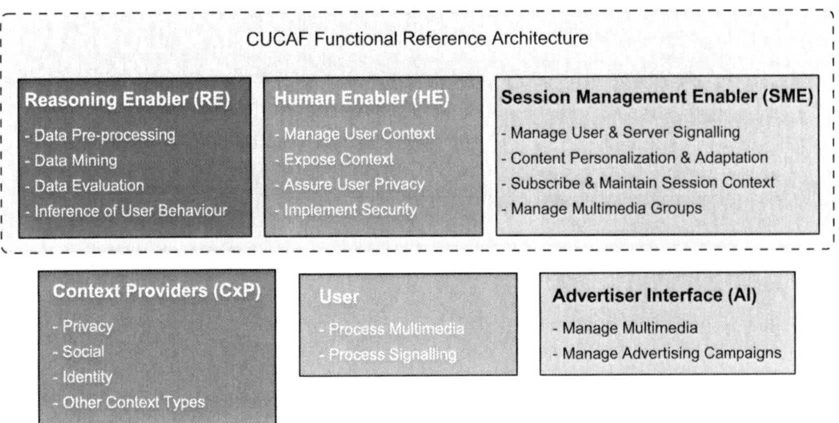

Figure 4.9 CUCAF functional reference architecture and adjacent entities

(Schulzrinne et al., 2003). In both of them, there are some interactions with a communication abstraction layer, which allows adaptation at the lower layers (e.g. network, transport). Finally, the *Context/Rule-Based Engine (CRBE)* is the module accounting for all the logic. In other words, it is responsible for subscribing or requesting context and trigger multimedia delivery accordingly. The advertiser possesses a direct interface with it, where the advertisements are converted into context subscriptions, rules and triggering points inside the CRBE.

The other functional entities, the *advertiser interface* and the *user*, are also important but are not the core elements of this architecture. The first represents an application – *Advertising Management* – (accessible via web), which allows advertisers to specify how each sort of user-related context will be used to influence the final advertising experience. The *Multimedia Manager* provides the interface for content to be associated with a specific advertising campaign. The latter represents the device of a person receiving some sort of advertisement. This can be a television, a radio, a mobile, a computer or any other multimedia capable device, which has both a *signalling module* and a *media stack*. The *context providers* represent entities that will constantly (and in real-time) provide updates about a particular context scope. Finally, the *identity provider (IdP)* is an external component that will be used as auxiliary to authenticate requests coming from the different advertising providers (advertisers). It is also important to notice that the CtPD does not belong to the SME, nevertheless the two components need to co-exist to allow multimedia delivery. A more detailed description about the capabilities and functionalities regarding the functional entities and respective modules can be found in Chapters 5–7.

Although a more accurate explanation regarding the interfaces between the different enablers is provided in the following chapters (5–8), Table 4.1 gives an overview of the type of information that is exchanged between the different entities.

Despite the argument for the enablers' approach has been explained in Chapter 1, after introducing them, the reasons become more obvious. The first resides on a one-to-one match with the technical sub-problems initially identified. This explains why the framework comprises three enablers and not another number. But probably the main motive lies on the possibility to reuse parts of the overall advertising framework to address specific needs. In other words and giving a

Table 4.1 Type of information exchanged between the different entities

	HE	**RE**	**SME**	**CxP**	**AI**	**User**
HE	–	Context	Context	Context	N/A	N/A
RE	Context	–	N/A	Context	N/A	N/A
SME	Context	N/A	–	N/A	Signalling	Signalling
CxP	Context	Context	N/A	–	N/A	Context
AI	N/A	N/A	Signaling	N/A	–	Multimedia
User	N/A	Signaling	N/A	N/A	Multimedia	–

concrete example, when a player can introduce new features at a reduced price, the business value of such solution is higher. Therefore, the enabler path may enable realizations of advertising architectures, which would not be sustainable when considering a complete system.

4.3.3 Realizations of an enhanced advertising system architecture

One of the design principles behind this architecture is its modularity and consequent flexibility. This is another key factor that influenced the decision regarding the enabler approach. The main idea relates with the possibility to place the different components into different domains and therefore originate new configurations that can bring improved business models or simply more efficiency, depending where and how the entities are placed and linked with each other, respectively. Assuming a simplified value chain (see Figure 4.10) scenario, this work defines three possible architectural configurations.

4.3.3.1 Independent enhancers and enablers architecture

This scenario represents an unstructured advertising environment, very similar to what usually happens nowadays. Each entity in the value chain is responsible for providing its own features through pluggable (to connect with other components) interfaces. Figure 4.11 represents the possibilities for distribution and composition of modules (enhancers) and components (enablers) among the different entities in the value chain, while Table 4.2 highlights the advantages vs. drawbacks of the different architectural scenarios.

The reason why the Advertiser Interface (AI) and the Reasoning Enabler (RE) always appear coupled with the Human Enabler (HE) relates with some dependencies between these entities. The first uses a user-profile XML scheme (see section 5.3) defined inside the HE, which is used to render the available context configuration possibilities. In what concerns the RE, the problem is the other way around, that is, the outcome of a prediction, or inference needs to be stored in the HE using a specific syntax (see section 6.4.3). Nevertheless, they can be used separately, as long as the interconnecting components respect these interfaces. Despite these comparisons, it is important to stress that these examples illustrate a mere architectural possibility.

In reality, there are other competences inherent to the presented entities (namely, media agency, publisher and operator), which cannot be replaced by simple pieces of software or hardware (e.g. a media agency provides the creativity that a publisher most of the times cannot offer). In fact, some of the architectural

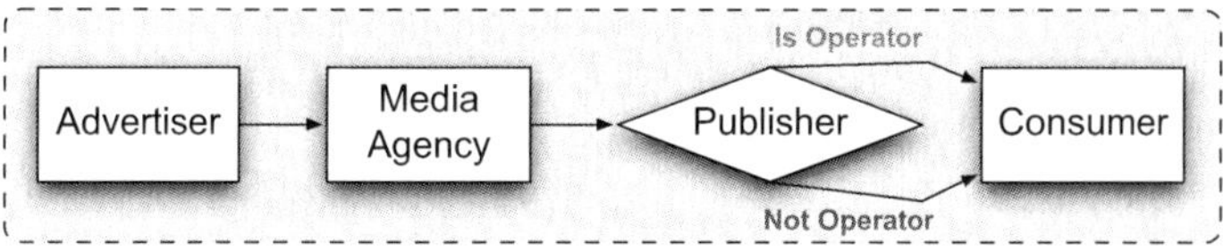

Figure 4.10 Simplified version of an advertising value of chain

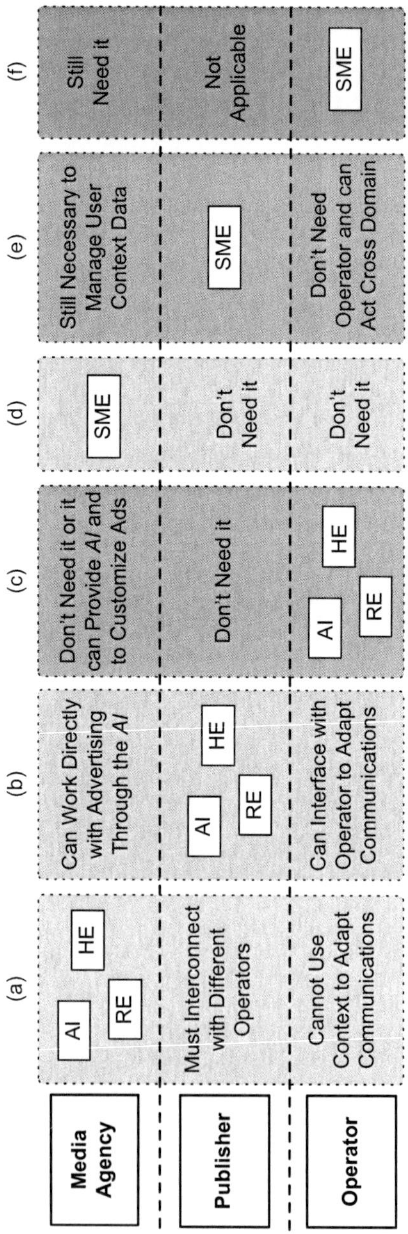

Figure 4.11 Scenario 1 – independent enhancers and enablers architecture

Table 4.2 Advantages vs. drawbacks of different architectural scenarios

Scenario	Advantages	Drawbacks
A	Media Agency can act as publisher when it has direct contact with operator or communications infrastructure. It would possess valuable information about people and could improve targeting mechanisms.	It is not likely that operators adapt their communication systems to context provided by single media agency. Publisher could act as broker but would increase complexity and introduce dependency.
B	Publisher doesn't need media agency. The AI allows them to address advertisers' needs directly.	It is more likely that operators can use context information but still difficult.
C	The operator can act as media agency and publisher. If it doesn't want this role, he/she can simply provide the AI so that media agencies can manage the relationship with the advertiser. Furthermore, operators can improve the knowledge about their users and their activities and consequently use this information to increase network efficiency and customer satisfaction.	In this situation all the interesting new functionalities for the advertiser/media agency are partially limited (they can still communicate with other operators but not influence adaptations based on context) to a single operator domain.
D	The media agency can act as a publisher and deliver content across a multitude of platforms and domains.	Despite having the possibility to deliver content, they don't know much about the users (except demographic information).
E	Just like in Scenario D, here is the publisher who can deliver content across a multitude of platforms and domains.	The same problems as in D, with the disadvantage of being dependent from the media agency information regarding the advertiser targeting decisions.
F	Operators can improve their infrastructure or better interconnect with other domains by abstracting from communication layers. Furthermore, they can use their own context information to improve network efficiency and customer satisfaction.	Might not have enough context information to take full advantage of the flexible adaptation mechanisms supported by the SME. Might be difficult to transpose all their functionalities to the media agency and therefore don't take full advantage of the SME.

configurations do not make sense at all (cases D and E). Furthermore, it is obvious that when combining all components, the outcome would be more beneficial for all entities; nevertheless, this scenario will be later explored (Scenario 3).

4.3.3.2 Full edge orchestrator architecture

The scenario where the full edge orchestrator architecture is in place is better suited for entities that want to offer an advertising service within a single domain. It enables players in the value chain to efficiently collaborate in creating, planning, executing, delivering and optimizing multi-channel advertising campaigns.

Furthermore, it provides reach and relevance via a single unified platform, with an attractive and easy-to-use graphical web interface. Figure 4.12 depicts the aforementioned architecture. Working under a single domain allows operators (or entities owning multimedia delivery infrastructure) to tap new revenue sources for their inventory (e.g. SMS or media gateways). By brokering external publishers' inventories, it also enables the system to expand its reach. In fact, the brokerage system can be used to interconnect with other operators, where in most cases, should correspond to an adaptation of already existent interconnections (i.e. they are used to change traffic between networks: SMS, voice, video, etc.).

In this situation, the system would use the HE to process all user-related data, the RE to infer new knowledge and the SME to handle the logic and the triggering of events/advertisements. Please notice that it does not use its delivery mechanisms, as in this case, the ones used are owned by the operator. Such architecture is therefore appropriate for entities already owning an infrastructure with existing relationships with other providers.

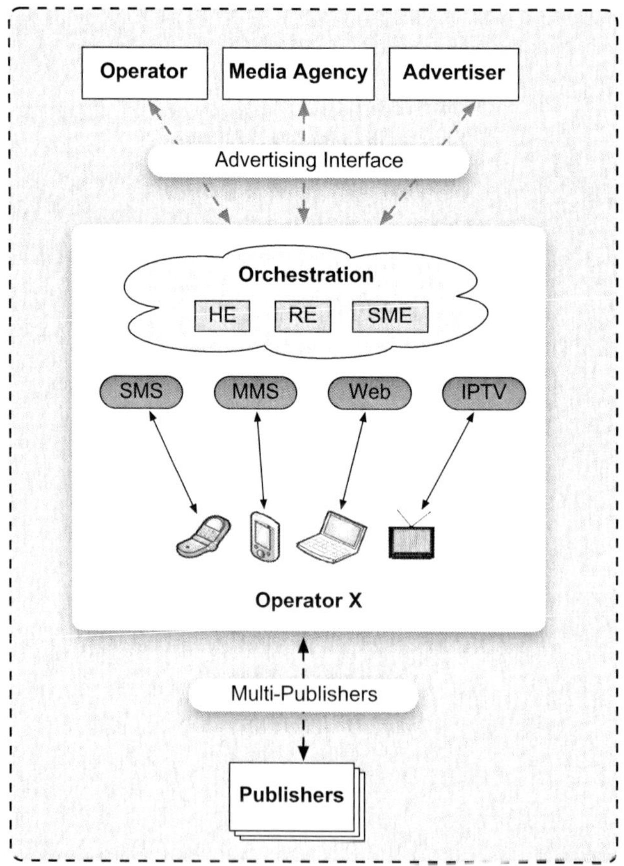

Figure 4.12 Scenario 2 – full edge orchestrator architecture

4.3.3.3 Full featured advertising solution architecture

This approach considers a full featured advertising solution, capable of acting cross domain, where all the components introduced in the reference architecture (see section 4.3.2) are used in their full capabilities.

Figure 4.13 exemplifies this scenario. The main difference from the previously presented Scenario 2 is that the SME is now responsible for all multimedia delivery across the different domains. These processes are further detailed in Chapter 7.

Obviously, the full featured version, as a broker solution, relies on good coverage of, and cooperation between, operators in the markets where the service is offered. And while this might be a difficult hurdle for some to clear, there are some entities around the world that regularly interact with the majority of the world's operators and have already established broker business relationships. Consequently, when this solution is leveraged by such players, the connectivity that is needed to jumpstart most markets is already in place.

4.3.4 Summary and contribution

This section showed how the CUCAF could be realized into an architecture and consequently a system, by introducing a series of complementary components.

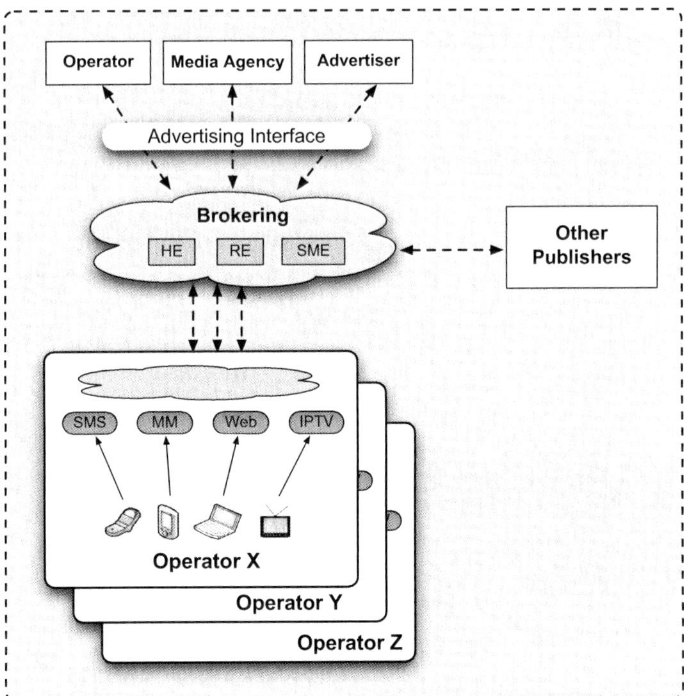

Figure 4.13 Scenario 3 – full featured advertising solution architecture

Moreover, it highlighted how the combination of the different entities can originate alternative system configurations. These, on the other hand, depicted how the same architecture can fulfil different (real-world) scenario's requirements. In fact, although there is no direct contribution towards standardization, the enablers have been designed to be compliant with existing standards. Also curious is the fact that the recent release of the Open Mobile Alliance (OMA) Mobile Advertiser Enabler specifications (http://www.openmobilealliance.org/technical/release_program/mobad_v1_0.aspx) resemble (in some ways) what is proposed within this work.

However, despite the utility of the proposed framework/architecture, it is important to notice that not all the components and interfaces were developed within this work. In this sense, this project contribution focuses mainly on the SME (and all the sub-components involved), the RE (and respective algorithms) and the AI. Nevertheless, this book further contributes with two context providers (further described in sections 5.5.1 and 5.5.2). In addition, it provides all the necessary interfaces to ensure security and privacy at the HE. Therefore, the client, the identity provider, some context providers, as well as, the CtPD entity are outside the scope of this work. Section 4.4 illustrates how each of the aforementioned components will be evaluated.

4.4 Evaluation criteria

In order to verify the benefits of the concepts developed in this work, evaluation metrics and methodologies are required. In this sense, this section provides some examples of how this can be done and describes how they can be used to verify the requirements defined in Chapter 3. The evaluation results can be found in Chapter 9.

4.4.1 Validation of functional requirements

Given the requirements specified in section 3.3, the best way to evaluate them is to make sure they were successfully implemented. Based on the amount of completed features, this system should be compared with other solutions presented on the related work of each main topic (Chapters 5–7). This will allow a better understanding of the relationship between this book and the remaining state-of-the-art. One example would be comparing the CUCAF with other user-context repositories, in what concerns the expected functionalities for such component (as the others are only focusing on a specific role, it is expected to outperform – or not – CUCAF). A complementary comparison between the proposed framework and other advertising systems is also provided.

4.4.2 Validation of non-functional requirements

As mentioned in section 3.4, the main requirements to evaluate are usability, performance, reliability and supportability. To understand the influence of each component within the proposed advertising solution, the evaluation is done independently, that is the metrics and methodologies are measured or applied, respectively, to each one of the four entities. Therefore, the components are evaluated in different (virtual) machines, enabling an independent control over the contribution of each one, towards the final advertising system. In this sense, the final evaluation is a weighted average of all components.

Depending on the component being evaluated, some metrics may or may not apply. Table 4.3 presents the relationships between the metrics and the components to be tested. Finally, to produce some valuable results and due to the impossibility of realizing tests with real individuals, most of the tests are performed by simulation tools. The exception goes to the metrics/parameters related with usability (which can only be evaluated by people). The description of these tests can be found in Chapter 9.

4.4.3 Validation of user requirements

Evaluating user-perceived QoE is the most important part of our validation work. Due to the impossibility of testing the entire advertising system together, in a real-world environment, the different components will be evaluated separately. While the validation of functional requirements will enable a better understanding of what is possible to achieve, the non-functional requirements will assess whether these tasks can be performed under reasonable limits, or not. Based on a combination of both, together with the requirements defined in section 3.1, a user survey should be performed. It should evaluate on a scale from 1 to 10 how users would feel by using the prototype implementation of the CUCAF. As an example, to evaluate the requirement ‘adaptation’, users can be asked if under a noisy environment, the replacement of audio with subtitles within a specific time limit (let’s say 2 s, which is the average time it takes for the CUCAF prototype) would be perceived as acceptable and desirable, or not. In addition, complementary questions to understand what would be expectable may also be included.

Table 4.3 Metrics to be evaluated for each component

Metrics	HE	RE	SM	AI
Affordance				x
Accessibility				x
Latency	x	x	x	
Throughput	x	x	x	
Efficiency	x	x	x	x
Scalability	x	x	x	
Fault tolerance	x	x	x	
Robustness	x	x	x	x
Security	x			x
Interoperability	x	x	x	x
Extensibility	x	x	x	x
Malleability	x	x	x	x

4.4.4 Validation of business partners' requirements

Although the validation of the business partners' requirements cannot be directly accomplished within this work (out of scope), it is somehow reflected inside some of the functional and non-functional requirements described in the aforementioned subsections. Therefore, Chapter 9 will introduce some of the business models that can emerge on the top of the CUCAF.

Chapter 5
Design and specification of the human enabler

As seen in the previous chapters, developing a system to tackle future advertising challenges is not an easy task. To simplify it, the approach taken within this work divided the problem into three complementary directions. The first, and the one being addressed in this chapter, deals with the human side of information technology systems. In other words, it covers human data representation and management, as well as privacy and security issues.

5.1 Introduction

Information and communication technology is becoming smarter, smaller and faster; and, at the same time, society is progressively becoming more densely connected. As a result, Internet-supported services are entering a new phase of mass deployment, which brings not only a huge number of new opportunities, but also new challenges. To address these challenges, the Future Internet (FI) needs to be tackled from a holistic perspective by taking into account all building blocks from users, services and applications down to the networks (European Future Internet Initiative, 2010). In this sense, it can be seen as a seamless fabric of classic networks and networked objects that will not only co-exist, but also be intimately bound up with our human world. It will be an Internet with Things, where the content and services it facilitates will be all around us, always on, everywhere and all the time (Hourcade et al., 2009). Therefore, using such an infrastructure, the advertising ecosystem will be no exception.

Despite all the technological revolutions, for the end-user (humans) it is the perceived QoE that counts, where QoE is a consequence of a user's internal state (e.g. predispositions, expectations, needs, motivation, mood, etc.), the characteristics of the designed system (e.g. usability, functionality, relevance, etc.) and the context (or the environment) within which the interaction occurs (e.g. social setting, meaningfulness of the activity, etc.) (Hassenzahl & Tractinsky, 2006). In other words, advertising must become personalized, contextualized, adapted, interactive, mobile, etc. while still concerning privacy. To achieve such scenario, it is critical to know more about the users. Consequently, it is mandatory to find a unified and standardized way of managing user data, that is accessing, storing, creating and modifying it. However, before focusing on the methodologies, protocols or technologies, it is important to understand what kind of data will leverage an optimized user experience in the advertising of the future. Hence, in a first instance, we propose a user identity data structure/profile that encompasses: user preferences, social networks and relationships, policies, devices, profiling algorithms, new knowledge generation, among others. Based on this data structure, a new architecture needs to be developed, enabling security, trust and privacy to be assured throughout the entire data management process. Altogether, the system should allow seamless interaction and integration with a myriad of things and services, independently from their associated technologies.

The remainder of this chapter is organized as follows. Section 5.2 outlines work related to user-data management, privacy, security, context-aware systems and associated challenges. Looking further, section 5.3 describes the proposed generic human profile concept, its taxonomy, and discusses the technologies involved. Based on the previously discussed principles, section 5.4 introduces the proposed architecture and explains how the system works. Chapters 8 and 9 complement the remaining work by presenting the implementation and evaluation results, respectively.

5.2 Related work

Current service creation trends in telecommunications and web worlds are showing the convergence towards a FI of user-centric services. In fact, some works (Baladron et al., 2009) already provide user-oriented creation/execution environments, but these are usually tied to specific scopes and still lack the capability to adapt to the heterogeneity of devices, technologies and the specificity of each individual user. Based on these limitations, the research in Liberal et al. (2009) identifies flexibility as the foundation for users' satisfaction, where the demand for different types of awareness needs to be present across the entire value chain of a service. Despite most initiatives require or propose some sort of user profile management systems, these are usually proprietary and include limited information about user preferences and contexts. Therefore, in order to make use of user information for a range of services and devices, there is a need for standardization of user-related data and the architecture that enables their interoperability. These efforts have been seen at both fixed and mobile worlds and are usually taken under the European Telecommunications Standards Institute (ETSI), the Third Generation Partnership Project (3GPP), Open Mobile Alliance (OMA), among others. Considering data requirements from a wide range of facilities and from different standardization organizations is the concept of Common Profile Storage (CPS) defined by 3GPP in tr 32.808 framework for user data for network services and management release 8. It represents a framework for streamlining service-independent user data and storing it under a single logical structure, in order to avoid duplications and data inconsistency. Being a logically centralized data storage, it can be mapped to physically distributed configurations and should allow data to be accessed in a standard format. Indeed, several approaches have been proposed to guarantee a certain interoperability degree and can be grouped into three main classes: syntactic, semantic and modelling approaches. The work in Bartolomeo and Kovacikova (2009) proposes their combination to enable interoperability of user profile data management for an FI.

Nevertheless, due to the pervasiveness of today's services, the amount of user-related information available is huge and dynamic. In this sense, user profiles should be somehow connected with such data. Context-aware systems are taking the first steps towards this direction. A number of architectures and models have been proposed and developed in the past decade to support the development of context-aware systems. The main differences were in the characteristics related to the application domain and techniques used for context modelling and dissemination. Initial projects focused only on localization systems and did not interprete contextual information, so they were not extensible or reusable. Pioneering this field was the Context Toolkit (Dey et al., 2001), a tool to aid the developers of context-aware systems. Its layered architecture permitted the separation of context acquisition, representation and adaptation processes. However, it had limited extensibility and did not support reasoning. Later, the CASS tool (Fahy & Clarke, 2004), a middleware for supporting the development of context-aware applications, provided a good abstraction of contextual information by using an object-oriented model for context description. The architecture specified a server containing a database of contextual information and a knowledge base with an inference engine. The mobile devices were equipped with various sensors to perceive context variation and send them to the server without local processing. Similarly, worked SOCAM (Gu et al., 2004), an architecture that uses the client/server model, where the context interpreter collects contextual information from context providers. The main strength of the SOCAM architecture is its context reasoning, which uses ontology for context description and allows a robust reasoning on context. Still, these projects were limited in their broadness and very limited to specific scopes. With a special focus on smart spaces, CoBrA (Chen, 2004) proposes an architecture based on a broker agent to support the development of context-aware applications in an intelligent space. The broker is an autonomous agent that manages and controls the context model of a specific domain. The broker agent contains: the context knowledge, context reasoning engine, context acquisition module and privacy management module. The broker agent collects context from devices, other agents and sensors of its surrounding environment and fuses them into a coherent model that is shared among devices and their corresponding agents. CoBrA uses ontologies for context description, which allows a good reasoning and a better sharing of contextual information.

Nonetheless, management is only one of the main tasks to accomplish within this ecosystem. User-data representation is a fundamental process as well. Information needs to be modelled for being machine interpretable and exchangeable using well-defined interfaces. The mains goals should be: easy manipulation, easy extension, scalability, efficient search and query access. In fact, the Strang and Linnhoff-Popien (2004) identify generic requirements: The modelling approach (1) should be able to cope with high dynamics, distributed processing and composition, (2) should allow for partial validation independently of complex interrelationships, (3) should enable rich expressiveness and formalisms for a shared understanding, (4) should indicate richness and quality of information, (5) must not assume completeness and unambiguousness and (6) should be applicable to existing infrastructures. Particularly important for historical information is that data is embedded with time constraints such as time of creation and validity. Recent large-scale projects that have explored context-awareness aspects include the European IST-FP6 Project SPICE (Service Platform for Innovative Communication Environment), OPUCE (Open Platform for User-Centric Service Creation and Execution) and MobiLife (Wireless world initiative). However, the proposed approaches fail to completely offer a generic, scalable and flexible architecture supporting both evolving context models and evolving services and applications. In essence, having carefully looked at strengths and weakness of existing context-aware systems, the C-CAST project (context to content casting) designed an architecture that aims to address the shortcomings of these systems and take advantage of recent developments in devices, sensors and techniques, and build up on the experience gained from earlier efforts in these domains.

However, standardization, interoperability, flexibility and management are not the only challenges. The research initiative in Sorge et al. (2009) shows that privacy, security and trust are major topics that deserve a special focus in what concerns user profile and identity management. Although usually tied to other IT fields, policies have been used to dynamically control the behaviour of systems for many different applications in both industry and academia. This trend reflects the benefits offered by policies, namely, efficiency, reusability, extensibility, protection and automation of routines in a system. More specifically, policy frameworks have been studied extensively in the last few years (Wright et al., 2002), being mainly used for network traffic management to control QoS access to data in the network and security. Several standardization organizations are involved in Next Generation Networks' (NGNs') standardization and each of them has designed policy frameworks to cope with management issues in the different levels of an NGN architecture. Particularly interesting is the work developed by the OMA, which specifies architectures to manage and control service enablers based on policies to authorize and enforce rules on third-party services. Within this architecture, the main component is the Policy Evaluation, Enforcement and Management (PEEM) from the open mobile alliance, which uses the Policy Expression Language (PEL) for that purpose.

There is also research in real-time policy languages and frameworks that address Service-Oriented Architectures (SOAs) (Phan et al., 2008). In an SOA system, all network, business and non-business functionalities are implemented as services, which are deployed independently of the underlying platform, normally characterized by dynamic changes (real-time service compositions) and heterogeneity. This complexity increases the difficulty to manage these systems, and therefore policies are considered as key enablers for SOA management. Many policy languages such as Ponder (Damianou, 2002), KAoS (Uszok et al., 2003) and

Rei (Rei, 2002) can be used to manage complex systems, offering not only a policy language, but also policy models and services.

In the industry, policies are being driven by many different factors originating from the necessity to offer new value-added services. As today's SDPs evolve towards becoming rich and enhanced, new service enablers and advanced devices increase the number of different deployment use cases, providing new constraints and requirements for the network and service providers in respect to QoE, as these solutions are developed at different levels of abstraction (e.g. network, application, service, business, etc.). Furthermore, they do not offer support for user-oriented policies, and therefore, systems cannot use people's desired configurations that best fit their needs within a determined context. 3GPP itself pointed problems with the fragmentation of their solutions towards a more dynamic and integrated environment, envisaging more control and flexibility for users through a study on policy solutions and enhancements. To address these limitations, user-centric services such as Facebook[1] are now giving users full control over their own experience, allowing them to configure how their data is accessed, shared and used. Nevertheless this is only valid within the application domain.

In this sense, based on the context-aware brokerage system (introduced in Chapter 4) and the proposed ContextML light-weight markup based scheme, this book addresses some of these limitations, offering an integrated system that considers, at the same time, both the user and operators/service providers needs. Furthermore it provides the interface that allows consumers to interact with the system, independently from the user-context type, the application requesting, or the entity owning it. Altogether, this work extends the framework presented in the C-CAST project by adding the necessary components. These changes will introduce the necessary security and privacy mechanisms and, consequently, position this work in the state-of-the-art of context-aware user-centric initiatives.

5.3 Generic human profile

5.3.1 Introduction

The Generic Human Profile (GHP) represents a set of properties built within a generic structure that allows the services of the future to use user-related information, while respecting their privacy, needs and concerns. Opening the door for opportunistic communications, user context is disclosed according to contextual privacy policies and settings, enabling systems and devices to sense how, where and why information and content are being accessed and respond accordingly. In addition, by using semantic, ontology and keyword technologies that understand the meaning of information and facilitate the accessibility and interconnection of content, it is possible to generate/infer new types of knowledge that can relate to users' behaviours, needs or intentions. Nevertheless, despite the utility of such

[1] http://www.facebook.com

profiling algorithms, the user should be in control during the entire process. People will wish to manage their identities in different ways, sometimes opting for full disclosure, and at other times disclosing only in an anonymous way that preserves their privacy. This is essential for establishing and managing trust and for safeguarding privacy, as well as for designing and implementing business security models and policies. By storing users external contexts, it will be possible to compare different sorts of data, so far not correlated, improving on the one hand the algorithms, but on the other hand, the user overall satisfaction as services become more contextualized, adapted and consequently personalized. Moreover, GHP envisages the integration of social data from different platforms, providing a unified way to access friends of users' (humans) lists, among others, combining both online and offline social network data. In the end, a crucial step will be the implementation of a Profile Description Framework (PDF) to handle the transformation of a technical profile into a tradable and interoperable good. In this sense, PDF should be at the heart of the GHP and other user-data management initiatives for advertising systems.

5.3.2 Requirements and technological considerations

When thinking about the implications towards future services, user-related data raise a series of questions. In what concerns security, it is necessary to consider a set of multi-factor authentication and authorization mechanisms capable of providing an open and standardized data management service. As for data storage and distribution, it is very likely to see trusted cloud service providers (probably Telcos) embracing this opportunity, where all attribute data related to users in a specific context can be easily accessed and aggregated. Although a distributed but interconnected standardized logic to access this data is required, its physical storage can be done within walled garden domains. In this sense, it is expected that context itself will have a signalling protocol and a complementary distribution one. Depending on the evolution of mobile devices, there is also the possibility to assist to a shift, where some of the users' associated data is stored locally. In fact, a first step towards the combination of the previously enunciated mechanisms (local vs. remote storage) is being explored by the Vodafone 360 service.[2] Regarding the business aspect, it is necessary that the solution is flexible enough to allow different entities to be involved in the value chain and therefore contribute to the overall service offering. Furthermore, billing and revenue distribution are also topics that need to be addressed, otherwise, when ignoring them, the overall solution may become compromised.

5.3.3 How is it represented?

In this framework, the Extensible Markup Language (XML)-based ContextML is used to model user-context information (aggregated within the GHP). Every

[2] http://www.theregister.co.uk/2011/10/18/vodafone_kills_360/

instance of context data is associated to a specific entity, that is every instance is the subject of interest (e.g. user, device or group of users), and comprises two elements: a type and a unique identifier. The type refers to the category of entities; exemplar entity types are username (for human users) and International Mobile Equipment Identity (IMEI) (for mobile devices). The entity identifier specifies a unique item in a set of entities belonging to the same type. Specific context information in ContextML is defined as scope, a set of closely related context parameters. Every context parameter is named and can be (1) a simple parameter, (2) an array of equal parameters or (3) a structure of different parameters. Parameters in of a specific scope are always requested, updated, provided and stored at the same instance of time. The scope design can be compared to object-oriented modelling, where the scope refers to an object class comprising various values as class members. For example consider scope 'position' which refers to the geographic location of an entity. This scope could comprise the attributes' latitude and longitude. The association between entity and scope is illustrated in Figure 5.1.

Whenever a context consumer requests or subscribes to a specific context scope, it receives a response encoded in the ContextML element 'ctxEl' when context is available. ctxEl contains information about where the context has been detected and encoded (context provider), which entity it is related to (entity), what scope it belongs to and the actual context data in the dataPart element. The elements par, parS and parA are simple constructs to store name-value pairs and attribute collections (structures and arrays), respectively. Every scope instance (context information) that is exchanged is tagged with a specific timestamp (time of context generation) and an expiry time. The expiry time tag states the validity of the context information. After this time, the information is considered invalid. Furthermore, a Quality of Context (QoC) value is added to summarize uncertainty and accuracy (Knappmeyer et al., 2010a). An exemplar ContextML snippet of the scope 'time' is depicted in Listing 5.1 (see Knappmeyer et al., 2009, 2010a, 2010b and the C-Cast project at http://www.ict-ccast.eu/for further details).

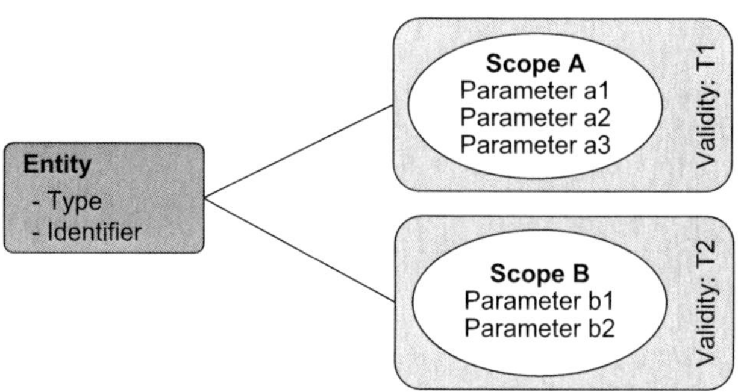

Figure 5.1 Relationship of entity and scopes

```
<ctxEls>
 <ctxEl>
  <contextProvider id="TimeCP" version=" 0.0.1 "/>
  <entity id=" john" type="username"/>
  <requestEntity id=" john" type="username"/>
  <scope>time</scope>
  <timestamp>2010-02-08T16:25:55+01:00</timestamp>
  <expires>2010-02-08T16:26:55+01:00</expires>
  <QoC>0.9</QoC>
  <dataPart>
   <par n=" timezone ">Europe/Berlin</par>
   <par n=" localTime ">16:25</par>
   <par n=" localDate ">2010-02-08</par>
   <par n="localDayOfWeek">Mon</par>
   <par n="weekend">0</par>
   <par n=" season ">winter</par>
   <parS n="dayTime">
    <par n="morning">0.00</par>
    <par n="noon">0.00</par>
    <par n=" afternoon ">1.00</par>
    <par n=" evening ">0.00</par>
    <par n=" night ">0.00</par>
   </parS>
  </dataPart>
 </ctxEl>
</ctxEls>
```

Listing 5.1 ContextML snippet of the scope 'time'.

5.3.4 What does it contain?

Despite the fact that the main characteristic of the GHP is its flexibility, this work defines basic types of information that will be extremely useful for a differentiated advertising experience. They can be either based on standards or defined according to specific service needs. These are depicted in Figure 5.2. In the following section, it is possible to find a brief description of the scopes and correspondent attributes defined for the advertising experience.

5.3.4.1 User profile and preferences

A user profile is a collection of personal data associated to a specific user. In this context, it refers to the explicit digital representation of someone's identity and stores the description of the characteristics, preferences and interests of a person. For this work purposes, it contains basic information about a user profile, as well as, other identifiers that can be further used to request user-specific context

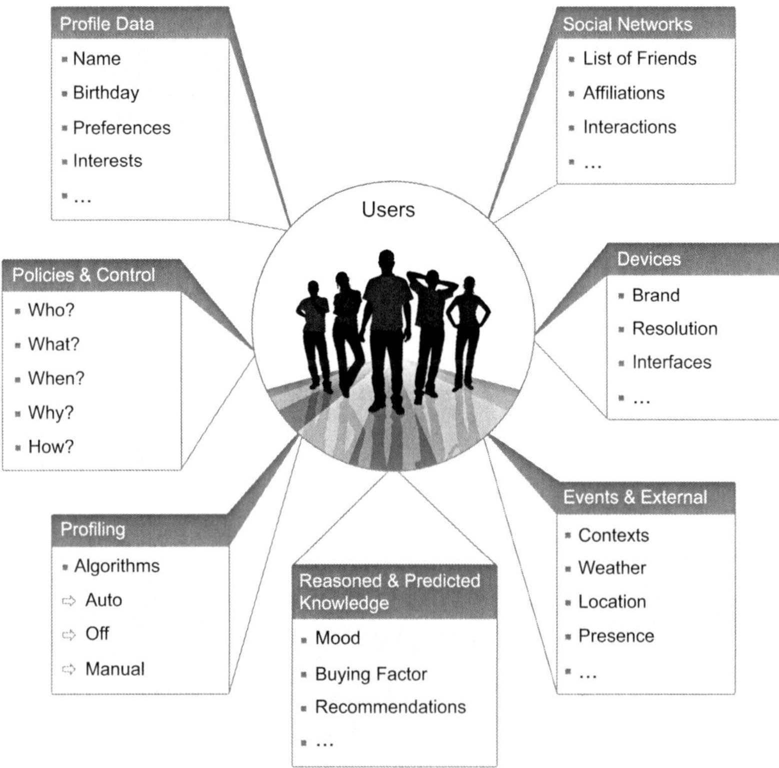

Figure 5.2 Example of a generic human profile taxonomy

(e.g. location of a specific user mobile device or list of friends of a specific social network). Table 5.1 contains the attributes being considered.

As mentioned earlier, attributes can be simple parameters or more complex object types. For the presented attributes, only 'addresses' and 'messengers' represent complex types. The first is decomposed into street, zip code, city and

Table 5.1 Existing parameters within the 'userProfile' scope

User profile		
• Username	• First Name	• Last name
• Gender	• Birthday	• E-mail
• Emei	• Phone	• Facebook ID
• LastFM ID	• Motto	• Addresses
• Messengers		

country. The second includes 'I Seek You' (ICQ), AOL Instant Messenger (AIM), Jabber, Microsoft Network (MSN) and Skype. Although the list of parameters can be extended, this list only describes the ones currently implemented.

5.3.4.2 Social networks

Humans in all cultures and at all times tend to form complex social networks. Despite in a real (offline) world these are hard to represent, the online variants offer a myriad of mechanisms to accomplish such task. In this sense, Table 5.2 shows, for each different social network, the list of attributes the GHP is capable of managing.

Not represented in the list are two extra parameters, the social enabler ID and specific network ID, which serve as identifiers for the social context provider and social network, respectively. Besides the basic profile information described in Table 5.2, there are more complex queries that can be performed to the social context provider. These will be further described in Section 5.5.2 and Chapter 6.

5.3.4.3 Devices

Due to the heterogeneity of existing devices, it is necessary to gather common characteristics among different models and brands. This information is captured by a context provider running on the terminal as a background application and gathers information about the different parameters. Table 5.3 summarizes some of them. This information is particularly important for session management (see Chapter 7), namely to assist initiation and modification operations. Nevertheless, applications for these data are uncountable.

It is also important to notice that the same context provider can support different scopes. In this sense, the different parameters can be distributed by two or more scopes provided by the Device CxP. This is particularly useful if a CxC is only interested in part of the information. As an example, for the session initiation process, the management entity does not need to know which network interfaces are available. On the other hand, the screen settings and media types supported are

Table 5.2 Existing parameters within the 'social' scope

	Social Networks	
• Religion	• Meeting sex	• Meeting for
• Relationship	• Relation with	• Political views
• Activities	• Interests	• Music
• TV	• Movies	• Books
• Quotes	• About me	• Home info
• Education history	• Work history	• Status
• Profile blurb	• Profile picture	• Last Update
• List of friends	• List of friends IDs	• List of groups
• List of groups IDs	• List of pages	• List of pages IDs
• Top influencing friends	• Top similar friends	• etc.

Table 5.3 Existing parameters within the 'devices' scopes

Devices		
• Ring mode	• Screen height	• Screen width
• Screen orientation	• Battery level	• Battery on charge
• Network operator	• Network type	• Service state
• Roaming	• Network state	• IP address
• WiFi MAC	• BT MAC	• Up time

of extreme importance. For this particular case, there should be a scope for the device network-related parameters and another for hardware and multimedia-related characteristics.

5.3.4.4 Context-awareness and events

There are several context types that can be associated to a user. Sometimes context can be acquired simply by providing the user identification (e.g. presence), while others might require extra context information to retrieve the desired data (e.g. weather needs location). Despite the variety of options/scopes available, this work focuses on location, weather, presence and calendar information. Tables 5.4–5.7 represent their attributes, respectively. Please notice that the amount of parameters associated to each scope can be changed at any point in time without affecting the overall functionalities of the system.

5.3.4.5 Policies and control

As initially mentioned, one of the main characteristics of the GHP is to allow the user to stay in control of all their related data. In this sense, this information is stored as policies in the PEEM component and replicated into the GHP in a ContextML format. Altogether, it allows the user to control who, what, when, why and how data is or can be written or accessed. The big difference towards other systems is that in here context information can be used as input for privacy settings. As all types of context can be used to create policies, it is not reasonable to cover all of them here. Instead, Table 5.8 shows some examples of what can be specified. Further details can be found in section 5.5.1 where the privacy context provider is introduced.

By default (when no policies are defined), all information is readable but not writable. Nevertheless, the system administrator may specify different default values for different scopes. When action 'read' is not specified and 'write' is, the

Table 5.4 Existing parameters within the 'location' scope

Location		
• Latitude	• Longitude	• Accuracy

Table 5.5 Existing parameters within the 'weather' scope

Weather		
• Current weather	• Forecasted weather	• Day
• Low temperature	• High temperature	• Humidity
• Wind speed	• Wind direction	• Summary

Table 5.6 Existing parameters within the 'presence' scope

Presence		
• Username	• First name	• Last name
• Gender	• Birthday	• E-mail
• Emei	• Phone	• Facebook ID
• LastFM ID	• Motto	• Addresses
• Messengers		

Table 5.7 Existing parameters within the 'calendar' scope

Calendar		
• Class	• Creation time	• Description
• Time start	• Time end	• Last modified
• Priority	• Sequence	• Summary
• User ID	• Transparency	• Attendees

Table 5.8 Examples of policies that can be set by the user

Type	Data	Condition	Action
Allow	Location	Is my friend	Read
Allow	Status	Belongs to the list of VIP applications	Read, write
Deny	Friends	Is not my friend	Read
Allow	Presence	Is my friend and requestor location is near me	Read
Deny	Location	My location = home	Write

system also assumes 'read' automatically. If, on the other hand, only 'read' is declared, the system does not assume 'write'. Furthermore, policies of type 'deny' have precedence over 'allow'.

5.3.4.6 Profiling

Within the proposed architecture, a series of different profiling algorithms can be inserted in the system. These, on the other hand, might be associated to a specific user application. Although somehow correlated with the policies and control, it allows the user to go one step beyond. This happens in a way, where the user is capable of controlling the profiling algorithm, choosing if it is 'on' or 'off', and if 'on', whether it runs on 'automatic' or 'manual' settings. This opens a new path for transparency, as users will know what kind of information is being used by each algorithm. Furthermore, it will give users the opportunity to choose which attributes/parameters they consider more relevant for their own profiling experience. Altogether, the same algorithm may have different behaviours based on the user self-defined profiling preferences (e.g. when inferring influence, some users may argue that the amount of friends in common is important, while others may indicate the number of photos together as a higher influence predictor).

5.3.4.7 Reasoned and predicted knowledge

With the information gathered from the profiling, the algorithms are usually used to reason or predict further knowledge about the users. These can be represented in a variety of forms, but concretely for advertising, it would be desirable if the user mood, willingness to buy, needs and intentions could be transformed in some sort of recommendation. However, these need to be provided by the user. Therefore, some inferred alternatives are desirable. The types of algorithms used in this work, as well as its details, are covered in Chapter 6.

5.4 Generic human enabler

In order to realize the vision previously described, it is necessary to transpose the generic human profile concept into a manageable format. In this sense, this section presents the architecture, the components needed, as well as some use cases, exemplifying how the system works.

5.4.1 General overview

With the goal of achieving an efficient and secure way of managing user-related data, the main concerns are:

- *Advanced data management*: allows the functions of data creation, modification, deletion, storage, acquisition, subscription, notification and syndication in a secure and efficient way.
- *Authentication*: act of confirming that someone is authentic and that the claims made by or about the subject are true.
- *Authorization*: function of specifying access rights to resources, according to a set of access policies.

To accomplish these objectives and to be compliant with the requisites specified in section 3.3.1, the architecture is mainly supported by two components, the privacy context provider and the human enabler, and its associated modules. Nevertheless, other entities may be added to tailor extra functionalities. Figure 5.3 represents the disposition of the involved elements. It is important to notice that all the different components can be placed in different domains without affecting any of the pre-enunciated functionalities. Before presenting how the system works, it is essential to understand which are the entities involved and what they do.

5.4.1.1 The human enabler

Acting as the main component of the entire architecture, the Human Enabler (HE) comprises four distinct but complementary modules that are logically coupled together, but can be physically separated (as long as basic security mechanisms are assured). They are:

- *Human context broker*: responsible for managing and processing all requests coming from the outside. Its interfaces allow direct access to recently cached information (context has always an expiration date) or an Application Programming Interface (API) for historical data. For the case of real-time context, it can be requested or subscribed. If it is requested or subscribed, when context changes, a notification is sent to the previously subscribed entity. Furthermore, it allows information to be updated, created or deleted using a specific ContextML format.
- *Human profile repository (HPR)*: where all the user-related information is stored (both real-time and historical). Basically it represents the physical implementation of the GHP specified earlier in section 5.3.
- *Policy evaluation and enforcement*: This component implements an interceptor that may be applied on all critical interfaces and act as an intermediate system between the resource requestor and target resource. It intercepts both the

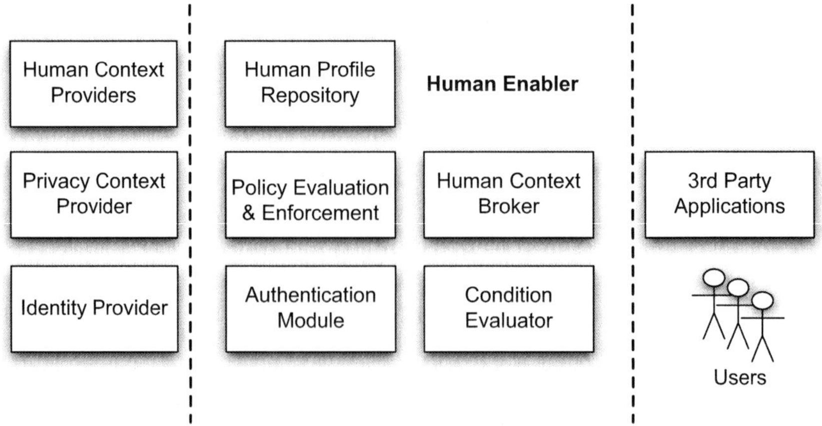

Figure 5.3 Overall human data management architecture

request and the response. Based on the evaluation results, this entity decides to forward the message to the destination or send a denial message to the message originator. The policy evaluation engine's main activity is the evaluation of policy conditions and execution of associated actions. In order to perform this activity, it has to first identify and return relevant policies from the policy repository based on the input data. The policy enforcement engine, of which the evaluation process is part of, builds an enforcement decision based on the results of the evaluation process execution and of the processes that imply invocation of other resource capabilities (request delegation) that are stipulated inside the evaluated policies.

- *Authentication module*: responsible for authenticating all requests coming from third-party applications or on behalf of other users. It verifies the authenticity of the token contained in the request.

5.4.1.2 Identity provider

An IdP allows users and third-party applications to come to a commonly agreed level of authentication for users and is able to produce the necessary formatting of authentication and authorization tokens. Even though self-asserted identity attributes will still be very prevalent in the GHP, there are also scenarios where the workflows will require token assertions of trusted attributes from identity providers. For an easy management of the roles or personas of users in this context, an identity provider will play the central role in such a user-centric setup. The identity providers in that sense can offer a secured life-cycle management of digital identities for users. However, this scenario assumes a relationship of trust between applications, domains and identity providers.

5.4.1.3 Human context providers

Human context providers are entities capable of obtaining basic or reasoned contexts from sensors, networks, devices, social networks or other data sources. Moreover, they provide and deliver this information in an interpretable manner, making it available to other components. They can act as standalone applications or publish their information into the HE through specific control mechanisms.

5.4.1.4 Privacy context provider

Privacy context provider is the entity responsible for controlling the user self-defined policies. It is responsible for synchronizing the policies between the PEEM and the HPR. This component is described in detail in section 5.5.1.

5.4.1.5 Third-party applications and users

Third-party applications and users represent the entities responsible for requesting user-related information. A third-party application can make requests on behalf of particular users or by the user itself.

5.4.2 Information management

As mentioned in section 5.2, one of the big challenges was finding a light-weight way to manage user-related data. Just like for the representation, the management is based on ContextML. Its schema is also applied for encoding management messages in order to allow for a flexible framework, supporting gradual plug and play extensibility and mobility. Furthermore, it was tailored to be used with REpresentational State Transfer (REST)-based communication between the framework components (namely the CxC, CxB and CxP). As user-context exchange always involves the Human Context Broker (HCB), it is important to understand what are its interfaces and what kind of communication types it offers. Figure 5.4 depicts the aforementioned. The content of the ContextML encoded messages is explained by an example in Knappmeyer et al. (2009, 2010a, 2010b and the C-Cast project at http://www.ict-ccast.eu/).

5.4.2.1 Context lookup service and advertising

Each context provider has to advertise its capabilities and how it can be accessed. Therefore, an advertisement message needs to be sent to the context broker. The broker stores the information in its lookup table. Every lookup entry expires automatically after a defined period of time (default is 2 minutes). Hence, each provider has to periodically renew its advertisement by sending a smaller Keep-Alive Advertisement message. This mechanism serves also for basic mobility, in case a component is deployed on a mobile device. If the IP address of a mobile

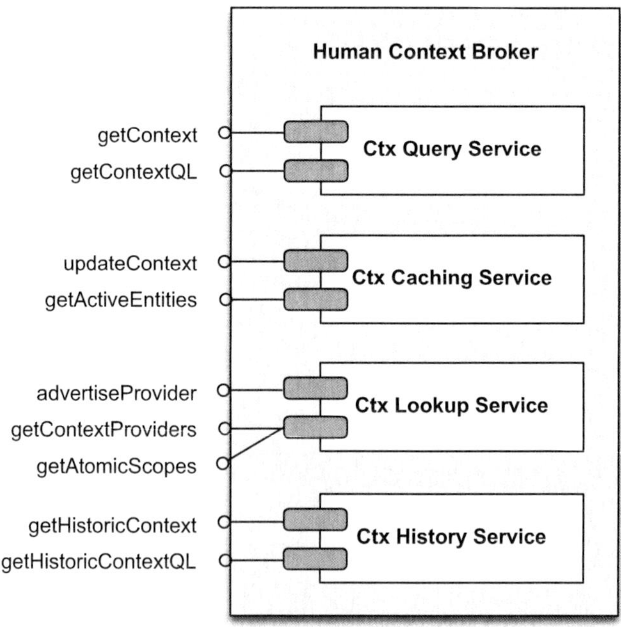

Figure 5.4 HCB interfaces' overview

provider changes, the consecutive advertisement will automatically fix the connectivity loss. Moreover, the advertisement procedure enables plug and play behaviour of the framework. The evolution of context-aware applications can easily be supported by adding new scopes during runtime – without having to redesign or restart the system from scratch. For this specific interface, the main functionalities are:

- *advertiseProvider*: method allowing a CxP to advertise its capability and been added to the lookup table of the human context broker.
- *getContextProviders*: method allowing a CxC or CxP to retrieve an entry of the lookup table.
- *getAvailableAtomicScopes*: method for retrieving the scopes that have been advertised.

5.4.2.2 Context caching service

This service is responsible for storing context into the cache, which contains all context information that has not yet expired. In this sense, it is used by CxPs to push context asynchronously to the HCB. Therefore, the HCB provides the following methods:

- *contextUpdate*: method for storing context into the cache of the HCB. This method is typically invoked by the CxP.
- *getActiveEntities*: method for retrieving all the active entities (i.e. context associated to these entities is available in cache).

It is important to notice that a context update is only accepted if the scope has been advertised previously by a CxP or if the scope is configured as primitive scope in the administrator settings.

Usually associated with a response, an Acknowledgment (ACK) is sent as a control message to confirm the reception and completeness of various management invocations (e.g. advertisement or contextUpdate). Besides the message, it also includes a numerical response code, mirroring HTTP status codes (e.g. 500 = internal error). Optionally and depending on the invoked method, the entity and scope information is encoded inside this message.

5.4.2.3 Context query service

The basic context query model is comparable to a key-value query concept. The triplet of entity identifier, entity type and scope serves as key, whereas the associated context refers to the value. Together with the access information, the required input data (i.e. instantiated context of a more primitive scope) has been communicated in the advertisement process. This circumstance highlights the distributed context processing and the layered abstraction. At the same time, it allows the following query methods:

- *getContext*: this method invokes the HCB proxy context query mechanism synchronously. The simple query model is used in this case (triple of entity id, entity type and scope). In the case of high-level context, the HCB fetches the required input context first.

- *getContextQL*: this method can be used for querying context using more sophisticated queries – advanced Context Query Language (CQL) – supporting both synchronous and asynchronous models.

When using the advanced CQL in synchronous mode, the SELECT keyword is required for selecting the action type. It supports the logical operators 'AND' and 'OR'. Regarding the comparison operators, the following are supported: EQ (equals), NEQ (does not equal), STW (starts with), NSTW (does not start with), ENW (ends with), NENW (does not end with), GT (greater than), LT (less than), CONT (contains) and NCONT (does not contain). For the asynchronous mode, an event based PUBLISH/SUBSCRIBE interface is supported, where CQL, which shows similarities to the well-known Structured Query Language (SQL), can be used to formulate the constraints, that is define the event of interest. Whenever the constraints match, a notification is sent to the subscribed consumers (CxC). For this option, it is necessary to provide a callback Uniform Resource Locator (URL), where notifications are sent in the form of a context update message.

```
[h!] [float=*] http://localhost:8080/CxB/getContext?
entity=username*scopeList=device
```

```
<contextQL xmlns=" http://ContextQL/1.1 "...>
  <ctxQuery>
    <action type="SELECT"/>
    <entity>username|*1 </entity>
    <scope>device</scope>
    <conds>
      <cond type="ONVALUE">
        <logical op="AND">
          <constraint param=" device.BTMAC" op="STW"
          value="8A:BF"/>
          <constraint param=" device. network" op="EQ"
          value="Simyo"/>
        </logical>
      </cond>
    </conds>
  </ctxQuery>
</contextQL>
```

Listing 5.2 (top) getContext vs. getContextQL (bottom)

All subscriptions are answered with an ACK message and contain the subscription identifier. As the initial subscription has limited validity, the component requesting the notifications is responsible for periodically sending a subscription renewal (containing the respective subscription identifier). When there is no further need for notifications, the requesting component can issue a subscription removal message to stop the notifications. Listing 5.2 illustrates both query mechanisms, as well as their particularities. The response for the first would contain the description of all devices belonging to all users. On the other hand, the second would return users' devices, whose network was 'Simyo' and base station MAC address started with '8A:BF'. This allows filtering at the CxB and, therefore, reduces the amount of information to be sent to the destination. In addition, it decreases the complexity of the application requesting it. It is important to notice that until this point, no security or privacy issues were covered. These will be introduced later in section 5.4.3.

5.4.2.4 Context history service

This service is very similar to the previously introduced. The main difference lies in the information validity. When using the Query Service, only context that is currently valid (existing in cache) can be retrieved. Nevertheless, for some applications might make sense to collect information from past events, this is particularly important for the work developed in Chapter 6. The other difference relates to the fact that PUBLISH/SUBSCRIBE mechanisms are not supported, as information is static and therefore makes no sense to implement such methods.

5.4.3 Privacy and security considerations

Assuming that the main functionality of such an architecture is to securely request user-related data (according to user contextual privacy disclosure), this subsection will demonstrate how the different components/modules communicate between themselves. Figures 5.5 and 5.6 illustrate these interactions. The architecture assumes that in the future, most applications have some sort of Internet-based dependency. In this sense, all user requests must be somehow authenticated (1). In this concrete scenario, the user is given the option to choose his/her favourite IdP. When he/she does this, his/her request is redirected accordingly (2). Afterwards the

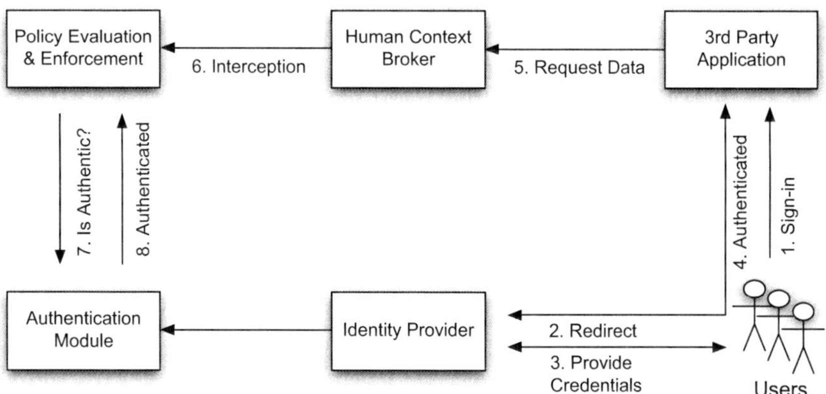

Figure 5.5 Authentication procedure for user-related data requests or subscriptions

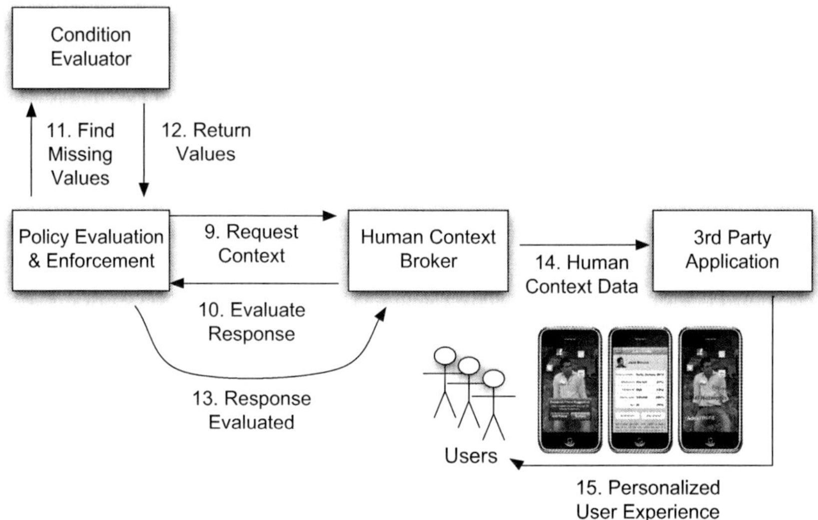

Figure 5.6 Authorization procedure for user-related data requests or subscriptions

user will authenticate according to the desired level of permission of the third-party application (3 and 4). The offered authentication methods can vary according to each IdP's capabilities for validating user identities. While offering tailored service towards users, applications might require extended user-related information (e.g. list of friends, location, weather, preferences, recommendations, etc.). Within the proposed solution, all requests are sent (5) to the HCB. Before being processed, these are intercepted (6) by the Policy Evaluation Enforcement & Management module (PEEM). Before checking or enforcing any type of policy, the component forwards the request (7) towards the internal authentication module so that the assertions provided by the third-party application can be cross-checked. In short, the token is verified by a Hash function. Once this is done, the PEEM is informed (8) and continues its operations. This allows the system to make sure third-party applications do not access unauthorized data on behalf of other users.

After the request is duly authenticated, two things can happen: if the message is a simple request, the answer will be processed immediately; however, if it corresponds to a subscription, the PEEM forwards the desired context information request (9); in other words, a pre-validated version of the initial request to the human context broker (before this step, the request is evaluated towards the provider/operator policies, stored inside the PEEM). When the message reaches the HCB, the broker stores the subscription constraints and triggers a notification when these are met. Therefore, independently from the scenario (simple request vs. subscription), only the response/notification is validated towards user policies (10). Because a simple request originates a prompt reply, the answer is automatically evaluated in the PEEM without the necessity of passing through the broker. In this case, Messages 9 and 10 are excluded from the scenario. Nevertheless, when considering a subscription request, they are mandatory. This happens because evaluating real-time context information at the time of the request could cause policy violation when the notification occurs (e.g. a user can specify that his/her social profile is only available when his current location is at least 1 km far from his office). When the PEEM starts evaluating user-related policies, it may happen that it requires real-time context data or the result of a conditional response to proceed with the validation. In this sense, the missing information is requested to the Condition Evaluator (CE), which is responsible for resolving the missing information (11). Once this is done, the results are sent back to the PEEM (12), which is then in the possession of all the information needed to validate the response. When finished, it is forwarded to the HCB (13), which uses the latest data stored in its cache to fill the necessary information. Finally, the response is sent (14) to its originator (the third-party application). Using the requested context data, together with the remaining application logic, the user is targeted with a personalized and adapted service experience (15).

Some policy examples, as well as a more detailed description of the security and privacy related issues, can be found section 5.5.1 and Appendix B.

5.5 Context providers

As mentioned before, context providers are standalone entities that can work together with the human context broker to extend the type and amount of information it provides. Each CxP has its own specificity and has to advertise its capabilities and inform how it can be accessed. Therefore, an advertisement message needs to be sent to the HCB. Then, at last, the CxP stores de the information in its lookup table (see Figure 5.4). Due to the fact that each lookup entry automatically expires after a pre-determined period, each provider has to periodically renew its advertisement by sending a small Keep-Alive message. The following sections show two examples of context providers implemented within this project.

5.5.1 Privacy context provider

5.5.1.1 General overview

When considering the amount of applications we interact with in our daily life, it is easy to understand that most of them are based on real-time systems. These are usually subject to operational deadlines and dependent on events to apply system responses. There are a myriad of applications that employ this philosophy and most of them are human oriented or provide some sort of benefit to them. Therefore, it is important to find a way to efficiently manage applications' behavior, in a flexible, but controllable way. In this sense, policies are typically applied to deliberate a plan of action, to guide decisions and to achieve rational outcomes.

Recently, with the increasing growth of user-centric services, to fulfil the basic user requirements, new management mechanisms are necessary. Consequently, it is important that systems can easily adapt to what people want, need and desire, without negatively influencing their personal experience. In this sense, there should be a way for policies to be applied to human context, while at the same time, giving users a way to control their own data.

Nevertheless, due to the complexity and nature of user-centric real-time systems, there are several challenges that need to be tackled. They are usually associated with the authentication and authorization of real-time data, which can be requested, subscribed or syndicated. To address the aforementioned issues, this work presents an architectural approach, an innovative policy syntax and a Privacy Context Provider (PCP), which, together, can be used to address specific user-centric scenarios in future real-time networked applications.

5.5.1.2 Architecture, interfaces and functionalities

Despite the simplicity of this component, it plays a key role in what concerns user-data privacy management. Basically, it acts as a broker between the HE and the PEEM entities and has the following main functionalities:

- *Profile control*: allows users to set which type of information (i.e. context scopes) is accessible to other users. Furthermore, it allows the specification of real-time context-aware policies (i.e. location scope is accessible if current location is different from home) for different context types.
- *Anonymization*: when activated, this option ensures that no personal information is used by the system or revealed to third-party providers. However, secondary data can be made available (e.g. location). Furthermore, when opting for this mode, the context information cannot be traced back to the original user. The users have the option to select which information is considered primary (e.g. name) and which is secondary (e.g. presence status).
- *Pseudonomization*: (from pseudonym) allows for the removal of an association with a data subject. It differs from anonymization (anonymous) in that it allows for data to be linked to the same person across multiple data records or information systems, without revealing the identity of the person. The technique is recognized as an important method for privacy protection of user health information (for more information see ISO 25237:2008 Health informatics – pseudonymization). It can be performed with or without the possibility of re-identifying the subject of the data (i.e. reversible or irreversible pseudonymization).

To better explain how the different components interact, Figure 5.7 represents different interfaces. In short, users can manage their settings using a direct interface with the PCP (I-1). These commands are then transformed into policies and saved into the PEEM (I-2). Optionally these can be replicated into the HE (I-3). Whenever a third-party entity wants to access or subscribe user-related data (I-4), the HE activates its internal evaluation mechanisms towards the PEEM. Further information regarding the way the policies are evaluated can be found in section 5.4.3. Therefore, independently from the type of privacy to consider, the way it is dealt by the system is always the same – through policies.

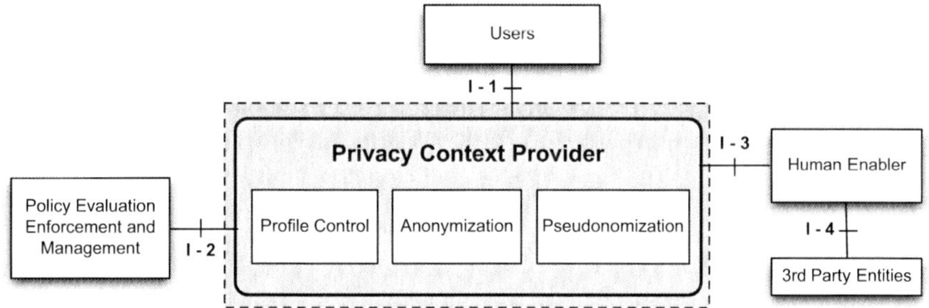

Figure 5.7 Distribution of the Privacy Enabler functionalities and interfacing entities

5.5.1.3 User privacy policies

To address the aforementioned situations, within the implementation of the PEEM, the Common Policy Format was chosen as a policy language (Schulzrinne et al., 2007) The main reason relates to its extensibility properties, making it suitable for the implementation requirements. Nevertheless, based on this work–specific requirements, some extensions were made to the current standard. These are further documented in Blum et al. (2009). To better understand how policies are defined, this section explains some of its attributes.

Conditions comprise two elements, which define the scope of the rule and the constraints imposed by it, respectively. The rules maintain the specified permissive character, which means only allowed message formats are represented. The scope of the rule describes in what conditions the rule applies, in order for the evaluation of the specific rule to proceed. Table 5.9 summarizes the main consisting elements (most of them were defined based on OMA recommendations to cope with the requirements previously specified).

The conditional evaluation process consists of two sequential steps, the validation of the rule scope against the input data and the evaluation of data constraints. The result of this process translates into a rejection or authorization of the message. The constraints section of the rule reduces the conflicts between policies at the definition level (see Table 5.10).

Table 5.9 Main rule scopes and their descriptions

Rule scope	Description
originatorIdentity	Requestor identity (address, name) of the processed message.
targetIdentity	Target identity under which policies must be evaluated.
Validity	Represents the time interval in which the rule is applicable.
Constraints	Similar to *sphere* but if they fail, the evaluation is aborted.
Sphere	Defines the prerequisites that has to be fulfilled in order to decide whether the rule applies and continue processing the constraints.

Table 5.10 Rule constraints and their descriptions

Rule constraint	Description
Operator	Describes an operator and its operands. In this case the evaluation type is boolean.
<and/>, <or/>	Represent the boolean operators, they are transformed in & and |, respectively, for evaluation.
<startDelimiter/>, <endDelimiter/>	Represent the parentheses left and right (e.g. (true&false)|true). These are used to increase the complexity of rules without needing to create new ones.
conditionalAction	Invokes a service capability and sets up constraints on the returned values. Its evaluation should also have a boolean result. The action must have a unique id inside the rule scope.

The actions component of a rule describes the activities that the PEEM has to execute further. Examples of activities are invocation of web service capabilities, transformation of the original message or forwarding the results to other delegations. Specific examples of user policies concerning anonymization, pseudonomization and real-time user profile control can be found in Annex B.

5.5.1.4 Innovations and use cases

Together with the PEEM and HE components, the privacy context provider can be used to improve user privacy in many different scenarios. In fact, it is particularly important because it allows users to set up policies, based on real-time data. This is valid for both user-context requests and subscriptions. Altogether, it can be applied for:

- *Collective intelligence*: one of the most relevant tasks for advertisers is to identify trends in specific communities, demographics, etc. By requesting anonymous data, user's context is not compromised and the important information is still retrieved.
- *Personalization*: probably the biggest challenge when dealing with privacy. By using pseudonomization, as opposed to user's context, user's identity is not disclosed, and, consequently, they can experience a tailored advertising offering.
- *Flexibility*: privacy parameters are no longer restricted to fixed values. Instead, they can be set to match any real-time context information.
- *Control*: being able to determine who can access what, at what time and under which conditions is a characteristic that should be available in any system that stores user-related information.

5.5.2 Social context provider

5.5.2.1 General overview

For more than 20 years we have been living in a society of online social relationships, where people tend to connect and share with others. Despite we have not yet seen its true utility, we are facing an era of social functionality, where social context can be used to personalize multimedia content and services. Altogether, when duly authorized and controlled by the users, the amount of available user's data might provide insightful knowledge that can be used to feed social experiences into next generation multimedia services. Using such information will enable services to become social-aware, connecting its customers into a more compelling experience and address some of the most important parameters that positively influence the overall user-perceived QoE, namely, personalization, contextualization, adaptation, interactivity and privacy.

Correlating these facts with research done around social networks, it is easy to identify parallelisms between what social communities offer and what users are looking for. These studies have been performed in Jamali and Abolhassani (2006) and Sorensen and Skouby (2008), where different aspects of social networks were analysed and matched with the user requirements. Moreover, social networks' adoption rate in the past few years confirms this trend (Owyang et al., 2009). However, due to the amount of existing social network websites, it is extremely difficult to extract information from each one of them in a standardized way that could be seamlessly used by different applications.

Contributing towards this vision, the following section presents an architectural approach that will enable multimedia social services, by securely exposing user social profiles, relationships and interests in a standardized way. Furthermore, it extends these data by providing interfaces to infer similarity and influence. In addition, by connecting with an external framework, it enhances the supported functionalities, allowing the calculation of relevant social network metrics. A special focus over the technological considerations, design decisions and innovative use cases is also included.

5.5.2.2 Architecture, interfaces and functionalities

With the purpose of enabling next generation social services, this work defines some goals, requirements and challenges:

- Sophisticated user social context and data exposure layer, where information can be requested or subscribed.
- Enhanced security, privacy and trust mechanisms, where entities requesting user-related context data are duly authenticated and authorized.
- Standardized interfaces to get information from different social communities.
- Provide a common framework that allows real-time social network metrics analysis and new knowledge inference: similarity, popularity and influence.

To accomplish these objectives and to be compliant with the social requisites specified in Chapter 3, this idea needs to be contextualized within the scope of the framework introduced in section 5.4 and the work presented in Chapter 6. These deal with privacy and security aspects of human-sensitive context information and try to understand and predict human behavior within next generation systems.

The Social Context Provider (SCP) is the entity responsible for providing social context information that can then be securely requested or subscribed according to user self-defined policies. Being inserted within this framework, the SCP can use the data mining and reasoning platforms to transform raw social data, acquired from the online social networks applications, into enriched information about a user's network and their peers. Figure 5.8 shows the entities and sub-components involved in these processes.

To allow flexibility, the SCP is decomposed into five distinct, but complementary, modules (technical details about the implementation can be found in section 8.2.4.3):

- *Social network application*: responsible for using the external APIs to request data about the users and their networks. This is done through the I-6 interface, which exists between the logic engine and every social network application. Furthermore, it represents the (visual) interface towards the end-user. Usually it is presented in some form of game or utility for the user.
- *Social logic engine*: contains the intelligence of the enabler. It accepts requests (through I-1 and I-7), processes queries (using I-1 and I-4), manages data internally (stores, deletes, accesses) and invokes metrics (I-9), as well as algorithm calculations (I-5). Moreover, it can act as a broker between the exposure module and the database (I-2 and I-5).
- *Internal database*: represents the physical storage of the social attributes, as well as the results from special operations executed by the logic engine. Moreover, it maps information from different social communities into a single logical entity – the user – whose data is fully accessible at the HE (by using the I-3 and I-8 interfaces). This data is periodically refreshed due to a Keep-Alive mechanism implemented on the logic engine, which informs the HE about its data structure. Based on such information, the HE updates its registries by invoking the respective web services (I-7). Alternatively, the HE might be

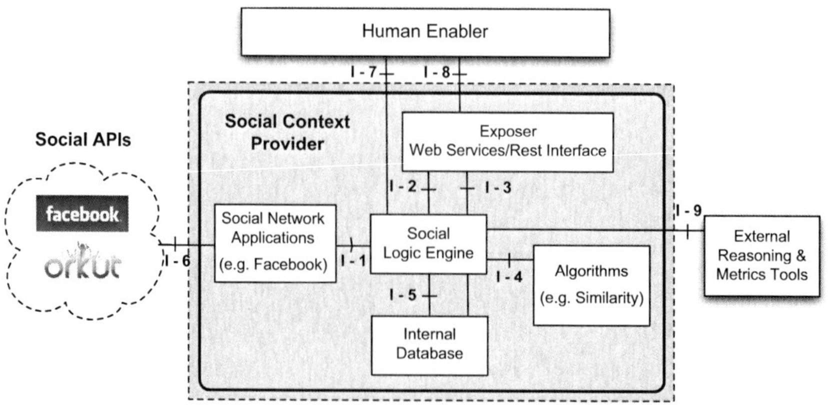

Figure 5.8 Distribution of the Social Enabler modules and associated entities

informed on every social context update. As all information has a validity, whenever it expires and there is an entity requesting it (directly or through the HE), it is automatically fetched from the SCP (I-8).

- *Algorithms*: to allow future extensions, a separate module, where a set of algorithms can be implemented without affecting the internal structure of the context provider, was created. These usually use raw social data stored inside the databases and request complex combinations of data before proceeding to the final calculation.
- *Exposer*: this module allows data to be queried by third-party entities. Within Converged User-Centric Advertising Framework (CUCAF), the only component capable of requesting information is the HE, through I-8. Then, internally, the requests can be translated into simple database queries (I-3) or more complex (I-2), requesting the 'help' of the logic engine.
- *External reasoning and metrics tools*: used to improve the type of information to be leveraged about user's social networks. They typically provide information regarding network metrics such as centralities, clustering, diameter, among others, which might be helpful indicators for external multimedia applications (see Chapter 6).

5.5.2.3 Design principles and technological considerations

Within this environment, when not contemplating the users and our system, there is an outside world that comprises both third-party entities and the social networks. Although it is interesting for social networks to expose their capabilities to external service providers, their regulations usually restrict them to run under a closed and supervised environment. Therefore, the same appealing opportunities that social networks leverage are excluding external multimedia services from using them. In this sense, most of the times, the user itself must directly fill its own personal information in order to be legal to store these data. Other restrictions also refer to a limited period of time under which the information may be stored (outside the Social Network (SN)), within the application. Furthermore, most of the information can only be requested when the user is using the application. This restriction creates the need for the applications to be both useful and catchy, making the user return and stay online for the time enough to collect all the necessary data, respectively. Otherwise, the SCP cannot address real-time scenarios with updated user information.

From a conceptual point of view, the SCP can be run by different entities, namely telecommunication providers, media and marketing agencies, the public sector, etc. In fact, Vodafone recently launched the Vodafone 360 service (http://www.theregister.co.uk/2011/10/18/vodafone_kills_360/) which integrates some of the presented concepts. Nevertheless, it is important to remember that due to the sensitivity of the data being collected and analysed, the entity responsible for controlling the Social Enabler (SE) must have a strong trust relationship with the customers/users. Another alternative, which is very often adopted within commercial products, is to offer rewards in exchange for the collected data. Either way, trust is a necessary requirement as information is not only internally used, but also exposed on request or subscription by third-party providers.

5.5.2.4 Innovations and use cases

Although the main goal of this provider is to improve the user-perceived QoE by using social data, until now, it was not explicit how this could be done. Therefore, the following use cases illustrate the applicability of such information within next generation multimedia services:

- *Recommendation systems*: knowing the similarity and influence of people with shared interests can improve content, product or service recommendations.
- *Group communications*: in recent context-aware services, it is common to group users with similar context information to improve personalization and efficiency.
- *Multimedia warehousing*: understanding how multimedia content is consumed, shared and interacted between users allows its owners to optimize storage and distribution mechanisms.
- *Advertising*: being one of the most promising revenue models, mining social data will help to understand what people want, need and desire. Furthermore, it will allow entering the era of 1-to-1 targeted advertising.
- *Network dimensioning*: seeing how peers interact and how the network evolves over time eases the prediction of future occurrences and consequently improves network planning efficiency.

5.6 Summary and outlook

With the increasing amount of information available on the Internet, managing and filtering it in an efficient way has become a major issue. Although many efforts are seen on the content industry (with sites such as Flickr, Facebook, Dropbox, etc.), the context part is often ignored or dealt within very specific scopes (e.g. most of the times, location). Furthermore, despite belonging to the users, it is usually hard to control context (e.g. delete it, share according to privacy settings) in an efficient way.

In this sense, by extending a previously known context brokerage framework, composed by a context broker, context providers and context consumers, this work provides a generic architecture for assuring privacy and security to any context-aware management system. This is achieved by combining the human context broker with the privacy context provider and a policy evaluation, enforcement and management system, resulting in the human enabler, capable of exposing user-related context to third-party providers. In addition, it provides ground steps towards a generic human profile structure, by providing a common way to describe user-related context. Moreover, with the deployment of the social context provider, this work exemplifies how the entire architecture can be used to facilitate access to one of the most important context sources in today's industry – social data.

Therefore, based on the aforementioned data, together with the architecture introduced, this chapter provides the basis for the remaining work. In other words, based on the developed infrastructure and collected information, Chapter 6 explores how user-related data can be used to infer new knowledge regarding their personal, social and environmental context. Likewise, Chapter 7 shows how the same system can be used to improve the balance between efficiency and personalization for next generation multimedia networked services.

Chapter 6

Design and specification of the reasoning enabler

When the challenges associated with the management of user-related data are dealt, it is important to take user's experience one step further. In this sense, understanding what users want, need or simply like is the obvious direction to pursue. Therefore, this chapter introduces a methodology which aims at improving the knowledge we have about a group of users, their interests and how these evolve over time. For this purpose, a special focus is given to the data collected from the social context provider (see section 5.5.2). In addition, it introduces some real-world applications for the generated data, as well as a standardized syntax (based on contextML) to expose this same information.

6.1 Introduction

Telecommunication and Internet services are constantly subject to changes, seeking the customer's full satisfaction. Improving these services with innovative approaches such as context-aware, social, mobile, adaptable and interactive mechanisms enables users to experience a variety of personalized services seamlessly across different platforms and technologies. However, to achieve such level, it is important to better understand their actions, interests and relationships.

Although the behaviour of individuals in online networks can be slightly different from that of the same individuals interacting in a more traditional social network (reality), it gives us invaluable insights on the people we are communicating with, how often this happens, which groups are we engaged with and which are our preferences. To overcome this discrepancy between online and 'offline' networks, data mining techniques can be empowered to approximate both worlds, providing awareness about people actual behaviour. They typically analyse different types of data to extract subtle patterns that may contain relevant information. The applicability of such studies can be seen in different industries. However, the most relevant are probably marketing and advertising, where identifying customer preferences, desires and needs is a key challenge in the conception and design of products and services. Furthermore, due to the variety of existing marketing strategies, it is important to identify the key peers in a network that will improve the efficiency of such approaches. Nevertheless, these two tasks cannot be seen as separate, as the most influential users for a concrete offering (let us say consumer electronics) are not necessarily the same ones for another (e.g. clothing).

In this sense, identifying influence and similarity within different communities of interest is still an unsolved challenge.

To understand how different user contexts can bundle together, collective intelligence can be applied. Basically, data is gathered from a large group of people, which allows drawing some statistical conclusions about the group that no individual member would know by themselves (Segaran, 2007). For the purpose of this study, the data being used was obtained from the Facebook social network. More concretely, the collected information includes: user's groups, pages, events, albums, photos, comments and tags, as well as personal information. Altogether, this set of contexts provided a great insight about users' preferences and interactions.

The remainder of this chapter is organized as follows. A concise survey of other initiatives to address social influence and preferences calculation is covered in section 6.2. A special focus over the concept, the data model and methodology behind this system is addressed in Section 6.3. Then, section 6.4 presents design decisions, lessons learned, as well as an abstract representation of a generic reasoning enabler framework. Afterwards, in section 6.5, some use-cases are shown, illustrating how this work could be applied to new business models.

6.2 Related work

The attempt of predicting human behaviour is not straightforward and involves a set of technologies that must be synergically applied to improve the estimation accuracy. Altogether, dealing with different aspects, they are capable of covering both the emotional and the rational aspects inherent to human behaviour. However, before getting deep into what has been done in these areas, it is important to introduce some basic concepts, terminologies and definitions that will be used throughout the remainder of the book. Furthermore, whenever relevant, these will be re-defined within the light of a specific work, application or context.

6.2.1 Concepts and definitions

One of the reasons why it is important to explore influence and user preferences within social networks relates to their ability to help understanding and predicting human behaviour. Although this term can have a myriad of definitions, in here, it is considered as an extrapolation of an individual's potential futures. Based on an interactive analysis process, a carefully crafted system using artificial intelligence, data mining processes and behaviour adaptive features can generate contextualized interactive personal future simulations in the form of interconnected micro-future scenarios, containing, alternative future paths and recommendations based on an individual's personal circumstances and environment (Koponen, 2010).

In this sense, being able to collect different types of user-related context became a priority for major mobile and web service/content providers. By user context this book refers to any kind of information that enhances an individual's awareness of the consequences of his/her existence, by connecting his/her personal and social actions to a wider social, cultural, political, economical and ecological (environmental) contexts. As this chapter will be dealing mainly with social data (also personal), for a better technical comprehension; to each dimension of this data (context), this work will reference to it as feature.

Regarding the term social network, we will make a distinction for three different perspectives. The first, personal, reflecting a specific user own network, comprises himself/herself and his/her first-level peers (nodes). A first-level peer denotes that there is a direct connection (edge) between the first user and a second one. The second type of network will be called global and refers to all the existing nodes and respective edges, in the system. The last one, particular, will refer to a sub-set of nodes and respective edges. This selection can be the outcome of a clustering operation, or, more simply, a specific customization for a special application.

Throughout this work, there will be some references to the term popularity of the user. Although its connotation is easy to follow, the term is ambiguous and therefore needs to be contextualized and re-defined according to the situation. Most of the times, it relates to popularity, similarity or influence, which are key topics underlying the understanding and prediction of human behaviour. In this sense, it is also important to present a definition for these social forces. Obviously, there are other forces due to confounding factors (e.g. external influence from elements in the environment), which can drive such metrics; nevertheless, they are outside the scope of this work.

A fundamental property of social networks is that people tend to have similar contexts (e.g. interests) to their friends. There are two underlying reasons for this. First, the process of social influence leads people to adopt behaviours exhibited by those they interact with; this effect is at work in many settings, where new ideas diffuse by word of mouth or imitation through a network of people (Strang & Soule, 1998). The second relates to the fact that people tend to form relationships with others who are already similar to them. This phenomenon, which is often termed selection, has a long history of study in sociology (McPherson et al., 2001), but is outside the scope of this research. Both the forces of social influence and selection are seen in a wide range of social settings: people decide to adopt activities based on the activities of the people they are currently interacting with; and people simultaneously form new interactions as a result of their existing activities (Crandall et al., 2008).

Therefore, influence, in its broad sense, relates to the ability of one person performing an action similar to his/her contacts. This can happen, either because more information is provided, or, simply due to the fact that it increases the value of the action to them. More precisely, given a particular 'action' A (e.g. buying a product, joining a community, etc.), an agent who performs A is called 'active'. Consequently, Bob has influence over Alice if Bob performing A causes or increases the likelihood that Alice performs A. In this sense, influence is a directed metric between two nodes. Although there are different types of similarity, the key concept is simple. In social psychology, similarity refers to how closely attitudes, values, interests and personalities match between people. In real systems, it can be seen in the form of opinions, interpersonal styles, amount of inter-communication, demographics and values (Larkey & Markman, 2005). However, generalizing it to a more mathematical concept, it can be seen as a correlation between two or more entities, where its value increases with the number of attributes these have in common. Unlike influence, similarity is an undirected metric, meaning that the value of similarity between two entities is always the same, although it can vary through time. Regarding popularity, it refers to the quality of being liked, common or having a high social status. This metric is a property of a single peer and is not associated to any link (edge).

To improve the readability and understanding of the work, all the aforementioned definitions will be further re-defined to accommodate the specific requirements of the proposed goals, methodologies and data. In addition, the following subsection presents previous research work performed on these areas, where the concepts introduced can be easily related to.

6.2.2 Influence

Especially over the last 10 years, there has been a great interest in online networks and communities. Indeed, with the rise of social networks, the activities within the research and industry communities have expanded significantly. Here a brief overview of what has been done in exploring preferences, interactions or relationships, how they can be measured over time and also where they can be applied to real-world applications will be provided. Furthermore, this section will introduce some of the methodologies, algorithms and techniques used by other works.

Initially, the work done within sociology and statistics has suffered from lack of data and focused almost exclusively on very small networks, typically analysed in the low lens of individuals. However, there is a long history of empirical work on this topic in sociology (Strang & Soule, 1998), through studies of effects, such as opinion formation and the diffusion of innovations. In economics (Young, 1998), theoretical models have been developed to cast social influence as a process, by which individuals in a social network tend to coordinate (or anti-coordinate) their decisions. Later, when data started to become available, computer scientists began developing models for influence in social networks, motivated by applications such as viral marketing. This was based on the premise that targeting a few key individuals may lead to strong 'word-of-mouth' effects, wherein friends recommend a product to their friends, who in turn recommend it to others, and so forth, creating a cascade of recommendations. In this way, decisions can spread through the network from a small set of initial adopters, to a potentially much larger group. Given a probabilistic model for the way in which individuals influence one another, the influence maximization problem consists in determining a set A of k individuals yielding the largest expected cascade. The influence maximization problem has been proposed and studied by Domingos and Richardson (2001) and Richardson and Domingos (2002) who gave heuristics for the problem in a very general descriptive model of influence propagation. Later Kempe et al. (2003) obtained provable performance guarantees for approximation algorithms in several simple, concrete, but extensively studied models from mathematical sociology. A continuation of the same work (Kempe et al., 2005) showed that the influence maximization problem can be approximated in a very general model (using a greedy algorithm) termed: the decreasing cascade model.

Based on the previously mentioned ideas, the work done in Leskovec et al. (2006) developed a scalable algorithm and set of techniques to illustrate the existence of cascades and to measure their frequencies. In their experiments, they found that: most cascades are small, but large bursts can occur; that cascade sizes approximately follow a heavy-tailed distribution; that the frequency of different cascade sub-graphs depends on the product type; and that these frequencies do not simply decrease monotonically for denser sub-graphs, but rather reflect more subtle features of the domain in which the recommendations are operating. This last finding is particularly relevant to the work this book is focusing on.

A slightly controversial study (Bakshy et al., 2011) states that 'ordinary influencers' are under many circumstances the more cost-effective alternative for cascading behaviours. However, they pointed out a curious fact: that cascading structures and behaviours change according to content or topic, reinstating the importance of understanding the domain being focused. Also interesting is the fact that all the research initiatives focused their work on collective behaviour analysis. More recently, the work from Cosley et al. (2010) considers two of the most fundamental definitions of influence – one based on a small set of 'snapshot' observations of a social network and the other based on detailed temporal dynamics – and studies the relationship between these two ways of measuring influence, in particular, establishing how to infer the more detailed temporal measure from the more readily observable snapshot measure. In their results, the correspondence between fine-grained ordinal data and its approximation made from snapshots is not perfect, but it appears close enough to make useful comparisons. This is particularly important for this work's research, as it was unable to retrieve all types of data in an ordinal manner. Despite all the advances done by previous work, very little is known on how influence relates with multi-context or multi-feature environments, because most studies were performed either at a theoretical level or within specific applications. Nevertheless, in the work by Crandall et al. (2008) some techniques are developed for identifying and modelling the interactions between social influence and similarity, using data from online communities, where both social interaction and changes in behaviour over time can be measured. Their results find clear feedback effects between the two factors, with rising similarity between two individuals serving, in aggregate, as an indicator of future interaction. They also consider the relative value of similarity and social influence in modelling future behaviour.

6.2.3 *Preferences*

Regarding preferences, research has been mostly focused on user interests and the relationships between each other – similarity. Trying to infer usability of user preferences for recommendation systems, the work developed in Lee and

Brusilovsky (2010) examines how similar are interests of users connected by self-defined relationships, in the collaborative tagging system *Citeulike*. Interest similarity was measured by the amount of items, meta-data and tags users shared between themselves. Their study shows that users connected by social networks exhibit significantly higher similarity at all explored levels (items, meta-data and tags) than non-connected users. This similarity is the highest for directly connected users and decreases with the increase in distance between users. Although it may seem logical, it is not necessarily true for all types of social networks. Another aspect is that preferences are very restrictive in such systems and rely on user self-defined tags, which are very likely to contain errors. Furthermore, it is very difficult to relate interests between people not using the same language. Focusing on a similar approach, but on a real-user social network (Facebook), the research (Bhattacharyya et al., 2010) developed a model to relate user profile keywords based on their semantic relationship and define similarity functions to quantify the similarity within a pair of users. Their concept introduced a 'forest model' to categorize keywords across multiple categorization trees and defined the notion of distance between keywords. Further, it used the keyword 'distance' to define a couple of similarity functions within a pair of users. Then, they analysed a Facebook dataset according to the model to determine the effect of homophily in online social networks. Although their results were similar to previous initiatives (direct friends share more interests/preferences), they made a striking observation: except for direct friends, similarity between users is approximately equal, irrespective of the topological distance between them. Again, this approach relies on user self-assigned data, which is very prone to errors. Furthermore, only a few of the users usually fill their entire profile. Heading towards this book's vision is the work developed by Spertus et al. (2005) which evaluates user communities (indirect interests) to understand similarity between users. They assert that just as we can estimate communities' similarity through common users, we can estimate users' similarity through common community memberships (e.g. user A might be similar to user B because they belong to *n* of the same communities). Besides being a not so computational expensive approach, it can even introduce better results like explained in Sarwar et al. (2001).

Another important metric within this area is popularity. Despite its broader concept, within our online society, it is usually associated with people or content (e.g. multimedia, products, services, etc.). The broad distributions of popularity and user activity on many social media sites were demonstrated to arise from simple macroscopic dynamical rules (Wilkinson, 2008). However, another approach (Hogg & Lerman, 2010) focuses instead on the microscopic dynamics, modelling how individual behaviour contributes to content popularity. Although examples from both concepts can be found in the literature, it is rare to find studies that examine the opposite direction, that is which people are popular among a specific topic, content or in a broader sense, interest. In fact, the existing works usually focus on a single dimension of data or feature. Therefore, it is necessary to introduce new methodologies that allow the exploration of multi-feature settings/environments.

6.2.4 Clustering

To allow a better understanding of how different types of data (features or dimensions) can be linked together, some approaches recur to clustering analysis. In short, this process represents the assignment of a set of observations (objects) into subsets (called clusters) so that observations in the same cluster are similar in some sense. An example of such idea was presented by Levine and Kurzban (2006). Although it focused on a single dimension (user relationships), they came out with interesting observations. It explained that clusters feature network externalities, which are not possible in sparse networks, thus conferring cascading benefits on the actors (peers or features) contained in those clusters. In this sense, complex implementations of such algorithm allow, not only the clustering of the objects, but also the features of the objects. That is, if the data (features) are represented in a data matrix, the rows and columns are clustered simultaneously. One example of such methodology is introduced by consensus clustering (Monti et al., 2003). In conjunction with resampling techniques, it provides a method to represent the consensus across multiple runs of a clustering algorithm (or different) and to assess the stability of the discovered clusters. The method can also be used to represent the consensus over multiple runs of a clustering algorithm with random restart, so as to account for its sensitivity to the initial conditions. Furthermore, it can be seen as a great approach to reconciling clustering information about the same dataset coming from different sources (Goder & Filkov, 2008).

This section presented some of the work being done around influence, preferences, popularity and similarity across social networks. In addition, it focused on some of its strengths and limitations to address the future of human behaviour understanding and prediction (by exploring social networks). Grounded on what has been presented, the following sections introduce some alternative paths of research by combining some of the aforementioned techniques.

6.3 The reasoning enabler

After introducing the motivation and some of the work that has been done, this section defines part of the research goals and the respective adopted methodology. In addition, it extends previously introduced concepts with analytical definitions.

6.3.1 Requirements and objectives

With the purpose of understanding and predicting human behaviour for improving next generation advertising, this work defines the following goals, requirements and challenges:

- *Discover the most popular users given a particular set of interests.* This analysis should be available from both a particular and a user personal network perspective.
- *Infer relationships between different user contexts.* This will help understanding what kind of user interests might be correlated.

- *Understand how user preferences vary and how influence spreads according to time.* This relates with the variation of user importance, within a previously identified set of interests.
- *Highlight the applicability of the previous points, towards an intelligent recommendation system.* This shows how the mix of influence and similarity (correlated interests) can increase its effectiveness.

Based on the aforementioned ideas, the following subsections describe the data model used to verify such conditions. Furthermore, it describes the methodology, as well as some of the design decisions implicit in this work.

6.3.2 *The data model*

One of the biggest challenges when working with data mining technologies is to find an appropriate dataset that allows scientists to make, or take, significant and meaningful inferences, or conclusions, respectively. The difficulty of such task is even higher when dealing with datasets obtained from real data, where the latter is usually not captured without errors. In addition, when the information contained within the dataset is related to humans, the sensitivity of the data raises complex privacy protection issues. Therefore, finding a dataset that allows scientists to study the previously presented (section 6.3.1) ideas is practically impossible.

In this sense, within this work, a new dataset was collected. To accomplish such task, a Facebook application that crawled user's personal information,

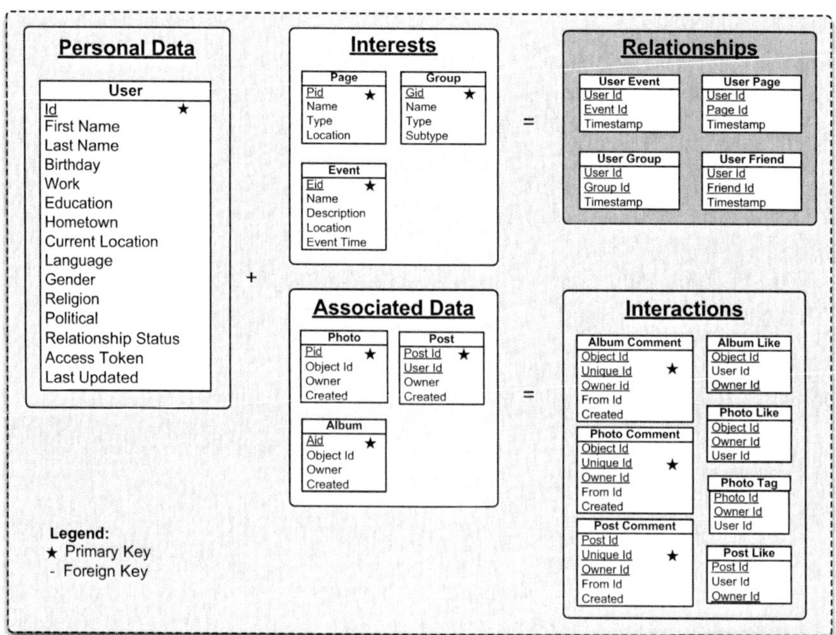

Figure 6.1 Representation of the data structure used for the collected dataset

relationships, interactions and related data was used. As seen before, with the rise of popularity among social networks, these tend to contain several pieces of relevant information regarding users. In fact, most of it reflects a very accurate representation of the users' interests, behaviours and demographics. An informal representation of the retrieved data, as well as its taxonomical categorization, can be found in Figure 6.1 (it also gives an informal illustration of the database structure). A more formal representation of the data can be found in the following subsections. All the data collected was done under strict user consent. All the security and privacy issues that may be raised regarding the type of information collected have been covered in Chapter 5.

6.3.2.1 Personal data

This type of data is a mere representation of user's demographics, simple preferences and orientations. This can be particularly useful to extract correlations with interests. For example, it may allow inferring that mostly male users between 18 and 25 are interested in football. However, its applicability can be further extended to other use-cases. Despite some of the attributes are directly filled by the user in a free text form (first and last names, work, education, hometown, current location, religion and political orientations), others are standardized by Facebook (id, birthday, language, gender, relationship status and access token). The field 'last updated' is used internally to indicate the last time the information regarding a specific user was refreshed. Although users are sometimes given the freedom to choose what to insert in certain fields (like seen above), the Facebook platform usually makes suggestions as they type, which increases greatly the uniformity of such data (e.g. location). However, there is always the drawback that data is sometimes entered in the user's preferred language, making the analysis of such information unusable for some applications.

6.3.2.2 Interests

Within Facebook, there are several alternatives to understand users' preferences and interests. In fact, there are specific fields in the user profiles where they are asked to input their interests, the activities they are engaged in music, books, TV shows or movies they like. However, these data contain a lot of noise. As seen before, by being free text-insertion fields, they can contain user self-made mistakes. As an example, if user John Smith mentions he likes 'Dr. House' show and user Maria Angelina indicated for the same field of interest 'Dr. Hose', their interest would probably not be correlated, when in real life, it should. Furthermore, these are often not filled in, or not accessible due to privacy settings.

On the other hand, Facebook allows users to associate themselves with groups, pages and events. Although the utilities of these properties are different, they usually establish a relationship between users and their interests. While groups are self-explanatory, where they represent a set of people with a particular interest, concern or affection, pages are usually used to connect people with their favourite brands, personalities, products, services or inclusively, groups. Events are adopted to organize gatherings, parties or meetings with user's friends, as well as let people in a community know about upcoming events. The main advantage of all these data is that it is standardized, and therefore, interests are easy to correlate between users. Moreover, each of them contains more precise information about users' interests, such as location, time or category (type and subtype). This level of detail can leverage the inference of new relationships not yet established between the different types of user preferences and interests.

6.3.2.3 Relationships

Relationships relate to the data resulting from the combination of users and their interests. This gives invaluable insight of who likes or engages with what. Furthermore, it gives the relationship between the different users. Because this metric analyses an undirected network (Facebook), the 'User Friend' relationship does not differentiate between the fields 'user_id' and 'friend_id'. Moreover, the 'timestamp' attribute does not relate to the time this relationship started or took place, but indicates the time it was crawled by the system. Although not optimal for a continuous analysis of data, this value can be interesting for snapshot analysis as explained later in section 6.3.3.4.

6.3.2.4 Related data

This kind of data is usually related with user-specific actions. In Facebook, this reflects the pictures and albums that a user owns/creates. At the same time, the content created by the user is stored under posts. This type of action can be an update to its status, a shared link, a video, a photo or a comment in a friend's wall (profile). Throughout this chapter, the aforementioned data types will be referred to as objects (albums, photos and posts). Although it was possible to get extended semantic information from such data (the content of the photos or posts), currently, the crawling only includes basic information that will help finding interactions between different users inside these objects.

6.3.2.5 Interactions

Interactions map 'related data' with the users themselves. It allows tracking which users communicate with each other and how they interact. In here, an analogy can be made with the 'relationships' introduced earlier. While the former shows which users are interested in which things, the latter illustrates which users are interested in each other. However, there is a big difference here. As opposed to 'relationships', 'interactions' give a directed indication. In fact, although classified above as 'related data', 'posts' are already a sort of interaction. Like before, semantic data could be collected to increase the quality of information (e.g. by knowing the content of the interaction, it would be possible see if it was positive, negative or neutral). Nevertheless, this is out of the scope of this work.

6.3.3 Methodology and algorithms

With the purpose of enabling the aforementioned objectives and based on the restrictions imposed by the previously presented data model, this subsection

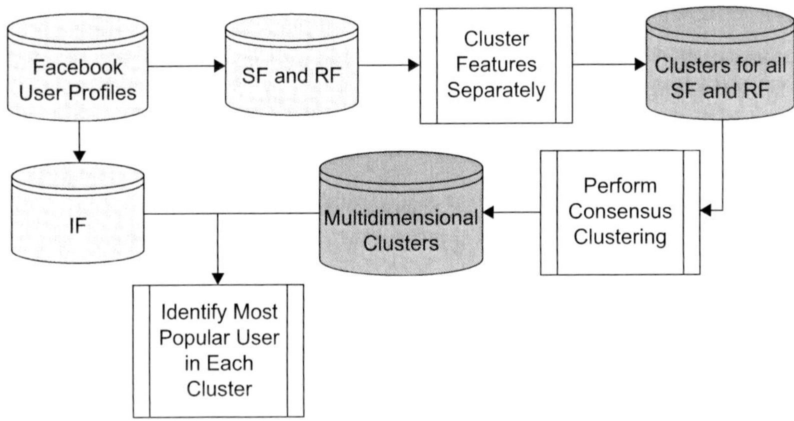

Figure 6.2 Methodology used in the system that was developed

introduces the methodology used within this research. Furthermore, it provides a detailed explanation of each of the steps involved in the reasoning enabler processes. In this sense, as depicted in Figure 6.2, the system comprises four main steps:

- One-dimension clustering (individual features).
- Multi-dimensional clustering (several features).
- Popularity calculation (based on interactions).
- Influence calculation (study how clusters evolve).

For a better understanding of the clustering processes and based on the data model described in section 6.3.2, the data types are sorted into three main categories. Each of them then comprises different types of features, which will initially be clustered separately, but later in synergy. *Stable features* relate to user personal information, which is rarely modified. *Dynamic features* are associated with users' interests. *Interaction features* include the relationships between the different user actions. Table 6.1 resumes the aforementioned. For the sake of simplicity, in these examples, it is assumed that the data is collected from the human enabler, then, cleaned, pre-processed, and only afterwards, used as an input for the data mining process.

Table 6.1 Categories of features used for clustering

Stable features	Dynamic features	Interaction features
Home location	Current location	User connections
Gender	Group membership	Common tags
Date of birth (age)	Events attended	Posts
Education history	Pages' association work history	Comments

6.3.3.1 One-dimension clustering

As previously mentioned, this research aims at analysing user's behaviour in order to predict their interests over time, as well as identify how these relate between themselves. To accomplish such task, in a first step, all features are analysed separately. There are two main reasons why this should be done:

- *Better data understanding*. It is important to have a notion about the structure of the selected features. This happens because data needs to be standardized for later stages of multi-dimensional clustering. For this purpose, text and numbers are considered as valid entries. In addition, it is important to assess whether or not it makes sense to cluster a specific feature. For data types such as 'gender', where only two values are expected, categorization may be more appropriate (as it is less expensive), where data is assigned to previously know categories (e.g. gender can be *male* or *female*). On the other hand, for features like 'age', 'group' or 'page membership', clustering is necessary.
- *Prepare features for multi-clustering*. Based on the outcome from each 'one-dimension' clustering process, data (cluster structures) can be uniformly used as input for the multi-clustering process. In other words, for each user feature vector, instead of using the actual feature value, the identifier of the cluster they belong to is used.

As an example, consider V_a and V_b to be features' vectors before and after 'one-dimension' clustering, respectively.

V_a = (home location = 'Lisbon, Portugal', gender = 'female', age = '25', education = 'Cornell University', current location = 'New York, USA', groups = 'Cosmopolitan magazine and MTV')

V_b = (home location = '0, 0, 0, 0, 1', gender = '1', age = '0, 0, 1, 0, 0, 0', education = '1, 0, 0, 0, 0, 0, 0', current location = '0, 1, 0, 0, 0, 0, 0', groups = '0, 1, 1, 0, 0, 0, 0, 0')

From the previous example, it is easy to understand the utility of the standardization process and consequent decision of using independent clustering for each feature. This stage is applied to both 'stable' and 'dynamic' feature categories. Only 'gender' is excluded from this operation, as it is categorized using the values '1' and '0', to identify 'male' and 'female', respectively.

Regarding the clustering technique to apply, a density-based algorithm, such as DBScan (Ester et al., 1996), is recommended. One of the reasons for this decision relates to the fact that this technique allows discovering clusters with arbitrary shape, and still achieves good efficiency. Furthermore, it does not require the specification of the desired number of clusters as an input parameter. As this method depends on a distance function to perform clustering, the decision went for Euclidian distance when the features were numbers and Manhattan distance otherwise (for text). Therefore, by varying some of the parameters (e.g. density threshold) used as input for the distance functions, it is possible to 'control' (influence) the way clusters are created (e.g. number of clusters, accuracy, etc.). In the end of this process, the system will obtain cluster structures for each feature and apply their feature vector as illustrated by V_b.

6.3.3.2 Multi-dimensional clustering

When all one-dimensional feature vectors are available, it is necessary to apply a specific technique to transform different cluster structures (each coming from its own feature) into a unified one. In this step, the previously (see section 6.2.2) described consensus clustering (Monti et al., 2003) methodology can be used. When this process finishes, the system produces clusters of mixed features, resulting in combinations of demographics and interests. This happens because some features relate to user demographics (e.g. age, location, gender, etc.), while others to their interests (e.g. groups, events, pages, etc.). To illustrate this step, Figure 6.3 exemplifies how single-feature cluster structures are transformed into a single cluster structure.

As a result, the final cluster structure allows understanding what kind of features and how their specific attributes can cluster together. In the example given in Figure 6.3, it is possible to see a correlation among users between 18 and 25, living in Portugal with an interest for nightlife. Similarly, users born in Germany are more propitious to be keen for technology. It is, however, important to notice that not all users born in Germany will like technology. However, the ones included in the cluster are likely to. Lastly, it is necessary to mention that the clustering structures in the upper part of the figure relate to features, while the multi-dimensional structure refers to users.

6.3.3.3 Popularity calculation

Once there is a comprehension of which interests and demographics group together, it is important to identify who are the key peers within these clusters. Although the term popularity can have different meanings, its connotation is usually associated to influence. In this sense, popularity is defined as a measurement of the user

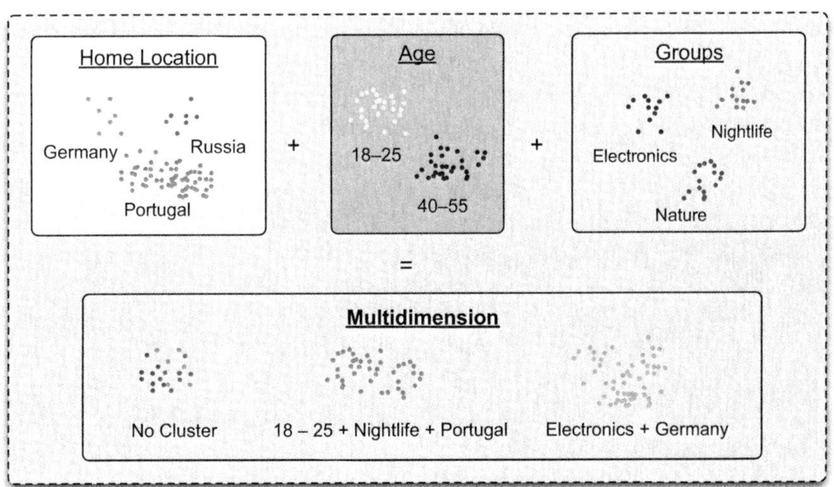

Figure 6.3 Exemplification of how multi-clustering process works

importance, within a cluster or a network (depending on the perspective), where importance is related to the number of interactions between the peers inside the cluster. In here, the term interaction is used to classify all activities that involve some sort of reciprocity between two users. For this case, the following attributes should be considered: common photos (users tagged together), common friends (connections), comments or posts directed to the user being analysed. Consequently, the popularity of a user u inside a cluster j is defined in (6.1), where i indicates an interaction type, $n_{i,u}$ the total number of interactions for user u, $n_{i,a}$ the total number of interactions for user a, T the total number of interactions, S the total number of users in a cluster, a, a user belonging to the cluster, L_i the total number of interactions from type i, and λ_i the weight given to interaction of type i.

$$P(u_j) = \sum_{i=0}^{T} \lambda_i \times \frac{n_{i,u}}{L_i} \quad \text{with} \quad L_i = \sum_{a=0}^{S} n_{i,a} \quad \text{and} \quad \sum_{0}^{T} \lambda_i = 1 \tag{6.1}$$

Alternatively, the term importance can be associated with the user position inside a specific cluster (e.g. metric based on cluster density and the relative distance between the peers). Another approach could be based on network theory analysis, where one could use different centralities to infer the nodes best located in the network/cluster. This is particularly interesting because different metrics could be used for different types of targeting. Betweenness is one example of a centrality measure that tries to find the central points in a network. However, in some cases, these values may be misleading measures of a node importance.

6.3.3.4 Influence calculation

In order to understand how influence changes through time and how it relates with a user's popularity, it is necessary to analyse how data changes within a determined time interval. Figure 6.4 depicts this process (for Case 1). It is done in the form of snapshots, where these represent an image of the dataset at a specific point in time, t. In this sense, for each snapshot, the following steps are executed: one-dimension and multi-dimensional clustering (sections 3.3.1 and 3.3.2, respectively), and, consequently, popularity calculation for each of the clusters identified. The set of most popular users in cluster j at a time t will be denominated as $P_{t,j}$.

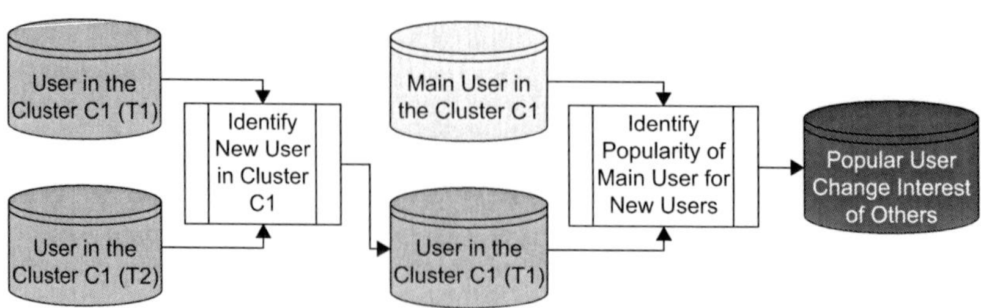

Figure 6.4 Case 1: How popularity affects influence within a cluster, through time

To proceed with influence calculations, it is necessary to identify how the user's popularity changed over time. Therefore, the first thing to be done is to make sure that it is possible to map the clusters from snapshot t_1 and t_2. As the resulting clusters contain users, this is used as a criterion to verify the similarity (i.e. the more users in common, the more similar they are). Afterwards, by applying the same methods as before, the set of most popular users for snapshot t_2 is calculated. At this point, it is possible to calculate influence.

At this stage, this work pretends to evaluate whether there is dependence between user's popularity (explanatory variable) within a group of interests or demographics and the influence (response variable) they have over other peers with similar features. In other words, it checks if popularity can be a good predictor for influence. In this sense, the following methodology is proposed to assess whether or not this dependence exists:

1. *New users in the cluster*: for each cluster, identify who are the new users (not existing in t_1 but present in t_2). Then, for each new user, compare the number of recent interactions (i.e. that occurred between t_1 and t_2) with all the other peers in the cluster. If the number of observed interactions is significantly bigger between the new users and the most popular users in t_2, than with the remaining users in the cluster, and this tendency is verified across the overall cluster structure (result of the multi-dimension clustering operation), it is possible to verify this dependency. Therefore, dependency is given by (6.2), where $n_{r,u}$ is the number of interactions between the new user r and another user u, and $L_{r,a}$ the total number of interactions between the new user and all the other users in the cluster.

$$D_{r,u} = \frac{n_{r,u}}{L_{r,a}} \quad \text{with} \quad L_{r,a} = \sum_{a=0}^{S} n_{r,a} \tag{6.2}$$

2. *Users no longer in the cluster*: if a user used to belong to a cluster in t_1, but it no longer does in t_2, then it is important to understand what happened. Did the user join a new cluster? If yes, repeat the methodology presented in (1). If not (meaning that it does not belong to any cluster), check if some of the users belonging to P_{t1} are no longer in P_{t2}. Again, if yes, repeat step (1) and check if the dependency between these is higher than the remaining users.
3. *Others*: the list of tests could continue, but instead of focusing on a test definition, which would evidence the observation (of how popularity can be used as a predictor for influence), this work gives some more examples of situations that could be explored.
   a. Evaluate if users follow popular users from one cluster to another over time (requires more snapshots of the dataset).
   b. Investigate if some cluster disappeared. In here, it would be interesting to notice if the most popular users in t_1 are still popular in t_2 but in another cluster. In addition, see how many of the remaining users followed this behaviour. This would show that popular users are capable of shifting user interests.

c. Check if new clusters were formed. In parallel, understand if the most popular users were previously popular (in t_0 or t_1) in any other cluster. This will show that besides the known influence types, popular users are also capable of setting trends or groups of interest.

As it is possible to see, independently from the observation in question, the same metric (dependence) is used to assert influence. The only thing that changes is the way data is analysed. Although it is widely known that metrics like correlation or dependence do not necessarily imply causation, it increases the degree of certainty in such belief. Furthermore, if it is possible to prove a dependency in all of the aforementioned methods (from 1 to 3), the likelihood that popularity can be used to infer influence is higher.

6.4 Design decisions vs. lessons learned

Concerning the reasoning enabler, although not conventional, this methodology focused on a set of specific steps, with the purpose of achieving some pre-defined goals (see section 6.3.1). This section, on the other hand, highlights some design decisions and proposes a generalization of the aforementioned mechanisms. The main reason why this is done relates to the fact that the context information available was not consistent across different context scopes. In this sense, while the previous approach focused on social context, this section uses the lessons learned to extrapolate a generic methodology.

6.4.1 The principles

When thinking about a generic framework capable of generating new knowledge, there is a set of criteria that should be met. Below is a list of the key design principles:

- Allow user behaviour modelling and prediction in real-time.
- Discover relationships and correlations between different context sources.
- Expose generated and inferred data in a standardized way.
- Provide a collaborative platform, where raw, simple or inferred context types can be shared and re-used by different entities.
- Enable all the previously mentioned aspects to be provided under a secure and controlled environment, where data can be created, updated, deleted, requested or subscribed.

To achieve the aforementioned, it is necessary to synergically combine the different technologies presented throughout this dissertation. In this sense, the next section presents the proposed generalization for a human-aware reasoning framework. Moreover, it covers the parallelisms between the approaches taken in section 6.3.

6.4.2 Generic reasoning framework

To enable the principles described earlier, this section introduces a generic reasoning methodology for human behaviour modelling and prediction. An overview of the concept is given in Figure 6.5. Although the exposure, control and, partially,

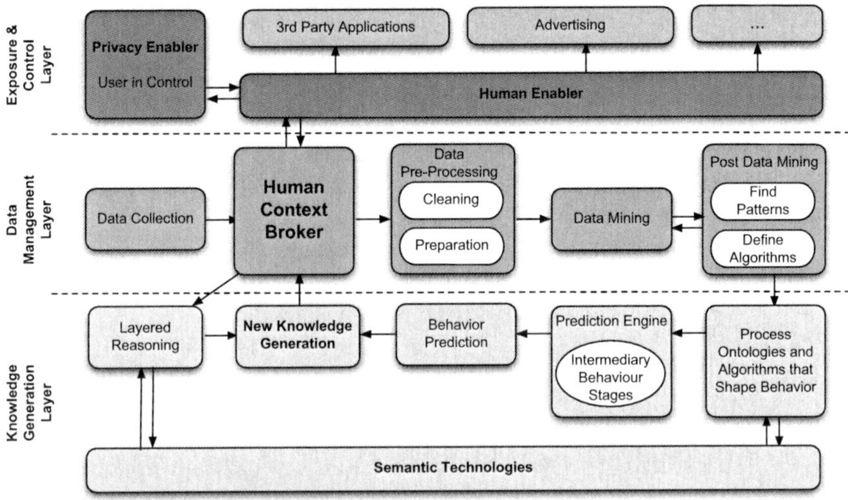

Figure 6.5 Generic framework for understanding and predicting human behaviour

the data management layers have been explained in Chapter 5, the remaining steps are new and will be further described, ahead.

6.4.2.1 Data management

This activity usually starts with data acquisition. This process involves gathering information from different systems. In here, any type of data can be considered, namely, user preferences, social networks' friends list, affiliations and interests, user devices' capabilities, as well as other external context-aware information (e.g. weather, smart spaces sensors, etc.). Afterwards, data is stored in a repository where it can be easily retrieved by any authorized entity. The next step is vital to the success of any data analysis operation.

Due to real systems' limitations, information is usually not captured without errors; therefore it is necessary to pre-process it in advance (before mining). Analysing data that has not been carefully screened for such problems can produce misleading results. Thus, the representation and quality of data are first and foremost before running an analysis. The pre-processing can be seen as a set of steps and includes data cleaning, normalization, transformation, denoising, feature extraction, selection, among others. The product of data pre-processing can be the final training set, or simply new data to be processed. In the previous case, it involved encoding operations (to normalize the data collected from Facebook), removal of invalid entries and selection of specific attributes and conditions (e.g. for snapshot analysis, it was necessary to select the same users and interests, otherwise the results would be distorted), just to mention a few.

Once this is done, data can be mined by using two different approaches: the first consists of using known statistical algorithms to help with pattern recognition and consequent algorithmic modelling. Despite the myriad of existing algorithms, they are commonly grouped into four classes of tasks (Fayyad et al., 1996):

- *Clustering*: is the task of discovering groups and structures in the data that are in some way or another 'similar', without using known structures in the data.
- *Classification*: is the task of generalizing a known structure to apply new data.
- *Regression*: attempts to find a function, which models the data with the least error.
- *Association rule learning*: searches for relationships between variables.

The second approach uses the opposite dynamics, where specific algorithms are designed to identify patterns and infer new data. As presented in section 6.3.3, in the previous approach, both methods were used. At the first stage, different types of clustering techniques were used (e.g. consensus clustering to find correlation between user interests), while at the second one, a set of algorithms were proposed (e.g. popularity, influence, etc.) to target specific outcomes. Although these last activities are contained within the data management sphere, they are already contributing to new knowledge generation.

6.4.2.2 New knowledge generation

At this point, by using simple context data, the system produces models that help understanding human behaviour. Based on the results obtained, new algorithms can be designed to infer new data. As seen before, the system cannot control the input data. Therefore, depending on the algorithms used to infer intermediary behaviour stages, it may be necessary to narrow or broaden the data source to accommodate with the algorithm requirements. In this sense, ontologies and semantic technologies can be used to achieve such goal. For example, a user might specify that he/she likes 'volleyball', a term that is imperceptible to the algorithm. Nevertheless, when using ontologies, this field can be transformed into 'sports'. Intermediary stages (or states) represent a set of inferences, which are usually associated with one's behaviour. These include mood, stress, receptiveness, social environment (e.g. influence), financial status, etc.

After these steps, and together with the previously modelled data, a new data mining process occurs – behaviour prediction. In theory, it aims at representing users' wishes, desires and needs. In reality, it presents a set of possible futures that could be interesting to the user. Figure 6.6 illustrates this process in more detail. Basically, it depicts a distributed human behaviour prediction model, which is based on different entities. In this sense, each company or entity can focus on their core know-how to predict, either intermediary, or final behaviour states.

Independently from the entity performing the predictions, the used context data (managed by the Human Context Broker (HCB)) can come from different sources, namely, raw context providers, the system self-defined data mining algorithms, or any other entity that tries to predict behaviours states. One important aspect to retain is the trust relationships between the different entities. When this affinity exists, the algorithms can use the accuracy of each context source in a weighted way (see section 6.4.3 for more details). This methodology reinforces scalability

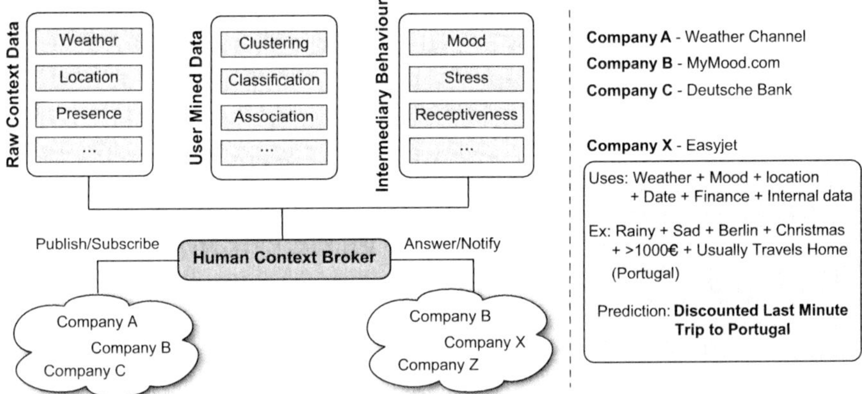

Figure 6.6 The process of human behaviour prediction

and performance, as the computational efforts, inherent to behaviour prediction, are performed outside the system. Only after, they are published to the HCB and made available through a third- party context provider.

6.4.3 New knowledge representation

To enable the collaborative scenario presented earlier, it is necessary that the reasoned knowledge is duly documented and represented. This way, third-party context providers will be able to re-use external data (coming from other entities), as an input for their algorithms. In this sense, it is necessary that all reasoned knowledge (as well as any other context source) is defined in a standardized way. Just like it was described in Chapter 5, ContextML is used as a descriptor. Listing 6.1 depicts an example of an intermediary state of human behaviour – influence.

In the example presented above, it is possible to see how influence can be described using contexML. In short, the message in Listing 6.1 represents the answer to a request issued by the user 'coca-cola', regarding user 'john'. It contains a description of the scope, the formula used to calculate influence (only if the CxP decides to share it), the calculated influence value for a particular group of users, the list of most influencing users, the list of users that he has influence over, as well as some of the basic features (i.e. demographics and interests) associated with this user. Furthermore, probably the most important field (which is mandatory in any scope), the QoC, indicates the confidence level of the CxP, over the information/ prediction provided.

6.5 Use-cases and business models

Although the main goal of this methodology is to improve the user-perceived QoE within next generation (advertising) services, by modelling and predicting user behaviour, until now it was not explicit how this could be done. The following use-cases illustrate the applicability of such system within next generation services and advertising systems.

```
<ctxEls>
  <ctxEl>
    <contextProvider id="ReasoningEnabler" version="
    0.1.2"/>
    <entity id="john" type="username"/>
    <requestEntity id="coca-cola" type="username"/>
    <scope>influence</scope>
    <timestamp>2011-04-08T16:25:55+01:00</timestamp>
    <expires>2011-05-08T16:26:55+01:00</expires>
    <QoC>0.67</QoC>
    <dataPart>
      <par n=" description ">Influence for user john
      </par>
      <par n=" formula ">secret (see previous
      sections)/par>
      <par n=" influence ">0.75</par>
      <par n=" influencerOf ">mtricko, pfizer, simon,
      maria</par>
      <par n=" influencedBy ">carlos, sofia</par>
      <parS n=" features ">
        <par n=" gender ">Male</par>
        <par n=" age ">24</par>
        <par n=" location ">Berlin, Germany</par>
        <par n=" education ">ISCTE, TU Berlin</par>
        <par n=" groups ">I love sports, Brazil,
        Volleyball</par>
        <par n=" pages ">coca-cola, MTV, hockey</par>
        <par n=" events ">1124, 7615, 6262</par>
      </parS>
    </dataPart>
  </ctxEl>
</ctxEls>
```

Listing 6.1 ContextML snippet of the scope 'influence'

6.5.1 Scenario A: advertising at the shopping centre

Today, most products or services are very well described in the digital world, and this information is usually available in a standardized way. Consequently, if there is a technical way of matching it with users' profile information, interests and current context, it is possible to improve the targeting accuracy and user satisfaction. Nevertheless, alone, this information is not very useful. A user may like sports, but it does not mean that he/she is interested in being advertised about it. The same concepts apply for other types of services.

This scenario exemplifies the situation where the user Elena enters a shopping centre and is presented with targeted advertising according to her tastes. She has the option to browse through the available offers, or be notified as she walks in the mall. The contents of the advertisements are of her interest and most of all, relevant, that is convenient. Suggestions are made based on what her friends bought or recommended, or simply by what other people that showed similar behaviour patterns (e.g. get inside the same shops, eat at the same restaurants, etc.). Furthermore, the advertisements can be based on real-time contexts. For example, assuming it is raining outside, Elena is presented with a discount coupon for the umbrellas' shop. A similar advertisement could be set for sun glasses, with inverse weather conditions.

6.5.1.1 The system side

But how can the system allow such a scenario? First, the triggering of the advertisement is only possible, either due to location or due to sensor technologies. Independently from how this happens, the system needs to collect and manage this data efficiently (at the human context broker). The same applies to the management of historical data. The clustering of people with similar patterns can be used to identify similarities with future user-context updates. By matching real-time context with previously mined data, the system is capable of suggesting something the user might want or need, or alternatively, identifies a candidate for a specific advertisement (e.g. a jewelry store should not be advertised if it is known the user recently acquired a ring). However, not only the users can benefit from this system, but also advertisers can now setup specific campaigns taking into account a myriad of user contexts (e.g. location, weather, time, social status and ultimately, behaviour).

6.5.1.2 The user side

Without the user control, the aforementioned scenario would be considered as a tremendous invasion of privacy. Nevertheless, based on the mechanisms explained in subsections 5.4.3 and 5.5.1, the user can express which entities can access each context type, at particular context conditions. In this sense, advertisers are also considered as 'users' within this system. Based on these premises, users are able to experience targeted advertising based on their interests. Furthermore, the same platform can be used to improve the multimedia delivery process as user device, network and environment context can be leveraged to third-party entities. This has been demonstrated in the course of this research work (Simoes & Magedanz, 2010; Simoes et al., 2009).

6.5.2 Scenario B: reduce churn, increase sales

Nowadays, internal directives within major Internet and telecommunication companies are setting churn reduction and sales increase as high-priority goals.

Although these companies usually possess information about how their customers use their services, they can hardly correlate it with other data. In this sense, correlating companies' private data with users' publicly available (e.g. limited to user privacy settings) information may open the doors for better estimations.

As an example, attributes derived from Social Network Analysis (SNA) may be used alone or as input to classical predictive models to help improve their accuracy. Furthermore, mining this information will help to understand what people want, need and desire. Therefore, marketing campaigns can be enriched by the intelligence gained through SNA capabilities. This is particularly important as even a small reduction in churn can mean big savings – the cost of retaining a customer is estimated to be only one-fifth that of acquiring one. Moreover, these customers could ultimately help decrease the churn within their own social circles. Additional marketing opportunities can even be derived from SNA for broadening the scope of existing customers, by identifying cross-sell and up-sell opportunities. In addition, the social status and position of key individuals in a network can have a profound effect on how their cohorts make decisions. Therefore, the ability to quickly and easily target those influential people can trickle scores of additional profit opportunities, across and beyond the entire contact circle.

6.5.2.1 The system side

Besides the possibilities presented in the previous scenario, it may happen that specific algorithmic efforts need to be performed outside the system. This can happen to protect entities core know-how, or simply to allow flexibility and scalability (see subsection 6.4.2.2 for details). Independently from the reason, the system must securely allow the update of user-inferred context by third-party applications. For this particular example, analysing a user's social network and running standard network metric calculations (e.g. betweenness, centrality, etc.) would allow identifying target members for a possible promotion. In what concerns the 'user side', the same principles apply as in the previous scenario.

6.6 Summary and outlook

With the uprising of user-centric trends, several service providers started to focus more on what users want, need and desire. Although these principles have always been true for the advertising industry, until recently, it was not possible to infer these. Nowadays, the amount of user-related information available opens the doors for human-behaviour understanding and prediction. Although technology is still far from perfection, under specific settings, it can be used to predict a possible alternative for a future event.

Taking some steps into this direction and based on data collected through a small set of 'snapshot' observations of a social network, the work done on this chapter tried to understand how users interests correlate (i.e. group of interests). Furthermore, considering a rich dataset of user preferences and interactions, by using clustering techniques, it identifies the most popular users within these groups. For this purpose, the strategy focused on multiple dimensions of user-related data, providing a more detailed process model of how influence spreads. In parallel, it studied the measurement of influence within the network, according to interest's dependencies. Based on the results obtained, a methodology was proposed to generalize the processes involved during the previously described (see section 6.3.3) strategy. Altogether, the work aimed at providing a generic way to understand the relationship between different user-related contexts by recurring essentially to data mining techniques and associated technologies.

Based on the work developed, advertisers, or any service provider willing to use this methodology and architecture (as specified in Chapter 5), will be able to infer new types of knowledge, previously not available. More concretely, within the scope of this work, it should be possible to setup an advertising campaign based on the most influential user for a specific set of interests. Although the effectiveness of such approach is not yet proved, this work provides an alternative to current targeting possibilities.

Chapter 7

Design and specification of the session management enabler

After understanding how user-related information can be securely acquired, managed and improved through reasoning and data mining techniques, it is important to explore how these data can be used to improve the overall multimedia delivery chain, within advertising scenarios. In a few words, this chapter describes a multiparty multimedia distribution framework, capable of balancing personalization with efficiency, across heterogeneous access networks. This represents the third step of our layered approach to address the complex world of advertising.

7.1 Introduction

Recent years, from about the early 2000s, have been characterized by global broadband penetration, fixed mobile convergence, triple play and content provisioning over all-IP networks, whether text, audio or video. Currently, the Telco world is challenged by the competition with the Internet and their quick time-to-market service provisioning. Indeed, the Internet domain challenges the Telco's business models massively, now aiming to catch the Web 2.0 opportunities. However, the goal is not to fight the Internet but to use similar Telco-enhanced assets in a greater extent, adding elements of adaptivity and context-awareness. For example using the user's context information can lead to context-aware content casting, group-oriented services (where groups are dynamically created) or context-aware adaptive multicasting (groups of people with similar preferences or requirements). Furthermore, when extending the user-related data to include network and environment context, the targeting and personalization possibilities become endless. Although the vision of this work aims at a personal advertising experience, the reality in what concerns targeting (mainly due to technical and economical reasons) is still limited to a set of users (groups) with common characteristics (context).

In parallel, with the increasing popularity of online community applications allowing users to form groups and exchange information in real time, multiparty communications are becoming widespread. In terms of group-based sessions, efficient multimedia delivery requires a correct definition of user groups. Nowadays, most mobile devices are produced with multi-homed capabilities, and it is common to cross areas where there exists overlapping of different network access

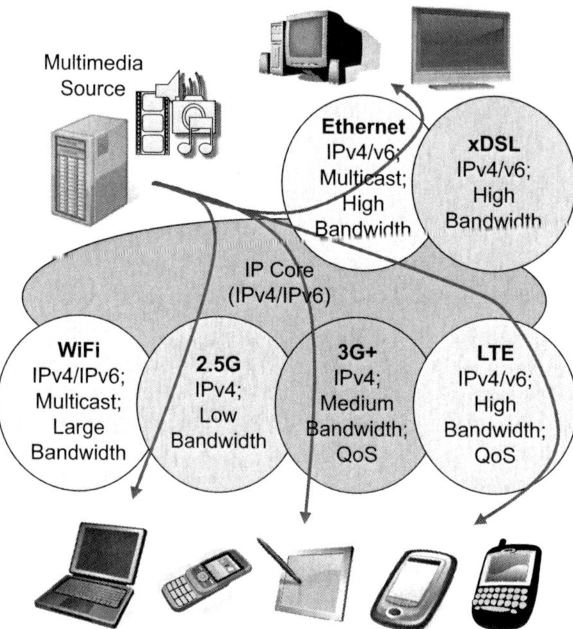

Figure 7.1 Different access networks that might be involved in the multimedia delivery process

technologies, such as WiFi, 3G and WiMax. Despite its convenience, the heterogeneity of networks and devices makes it difficult to roam between them in a seamless and non-intrusive way. Figure 7.1 exemplifies some of the possible access networks, basic characteristics and type of devices usually associated with them.

Therefore, the efficiency of the grouping operation (creation of a set of users to receive a given session) may depend on a few parameters, including the access technology. For instance, 3G networks have lower bandwidth capabilities in comparison to Wi-Fi and WiMax networks; thus, depending on network conditions, for 3G users to be able to join a session, capacities should be carefully considered. Thus, sub-grouping could be performed and the same service session could be delivered with different throughputs (e.g.– using different codings of the same content) to adapt the throughput to the current network capabilities. In addition to network traffic, other types of context information should also be used to improve sub-grouping. To tackle these issues adequately, it is necessary to consider optimizations at different layers, namely, application, session, network and transport.

With the aim of addressing these issues, this work touches all of the aforementioned topics. The application layer will be handled by the advertising interfaces. The session layer, thoroughly covered in this chapter, will be dealing with all user-to-content and content-to-user relationships, by providing all the necessary signalling to deliver content to its consumers and handling different types of events. Although not emphasizing the network and transport layers, this work will show

how it can interface with them to improve the overall user-perceived QoE within multiparty advertising scenarios. In this sense, the rest of this chapter is organized as follows. Section 7.2 introduces work related with multiparty multimedia delivery systems. Later, section 7.3 describes the overall context-aware multimedia management system, namely its architecture and components involved. The focus of section 7.4 goes to the main component of the system, the session management enabler. Finally, some practical examples and functionalities are exposed in section 7.5.

7.2 Related work

With the increasing popularity of online community applications allowing users to form groups and exchange information in real time, multiparty communications are becoming widespread. One application example is the group-based real-time video streaming service, TVperso (http://www.free.fr/adsl/pages/television/tv-perso.html), offered by Free, a French Internet Service Provider. This service is supported by IP multicast and allows users to stream their own content in real time, either publicly or to a limited group of receivers. From the research world, the work developed by Luo et al. (2007) presents a multiparty videoconferencing system based on a P2P solution. Identifying the heterogeneity of the Internet users as the biggest challenge proposes an application-level multicast scheme. The main goals are to adapt multimedia sessions to the types of Internet connections (direct connection to the Internet or behind firewalls and Network Address Translation (NAT)), to the amount of governable network resources (e.g. available bandwidth), and to the requirements of software features. Among other features, they support both multicast and unicast connections. However, this work was not validated within mobile environments. Providing a complete content production and delivery end-to-end solution, the ENAMORADO architecture (Papagianni et al., 2008) targets interactive scenarios, where the real-time monitoring of network conditions allows the appropriate selection of the content version and optimal usage of network resources. Despite the improvements, the framework is limited to the network context and cannot adapt the session once this is already created.

From the literature, context-awareness has been identified as a key enabler for enriching/optimizing group applications, allowing various forms of dynamic adaptation of a service, based on the group members' context (Coutand et al., 2005). Typical enhancements proposed include the dynamic formation of a group based on users' context (e.g. location), or the dynamic, preference-based selection of the (same) content received by group members using group preference arbitration systems. Moreover, personalized sessions can also be influenced by varying context, thus allowing users to join sessions based on their indoor location, preferences, profile and capabilities (Coutand et al., 2005). In fact, different communication behaviours, such as user-initiated communication mediated by the network, situation-oriented and network-initiated communication mediated by the environment, will be possible, when the concept of smart spaces becomes an integral part of global group communications. In this sense, technology should aim at extending the smart space concept in heterogeneous networking environments to provide ubiquitous computing, support of dynamic group management and personalization.

Taking advantage of such vision, some research conducted (Handorean et al., 2005) proposes context-aware session management for services in ad-hoc networks. It exploits strategies involving the use of contextual information, strong process migration, context-sensitive binding and location agnostic communication protocols to offer follow-me sessions. Another alternative is presented in Qian et al. (2009), where a service-oriented multimedia delivery system for pervasive spaces is introduced. To address the heterogeneity of devices and connections, it proposes the generation of several functional paths that can meet users' requirements. Therefore, based on context information, the system should choose the most suitable media quality delivery path. Furthermore, aiming at always providing the lowest delay, the system allows the change of data volume during the data delivery (i.e. by changing the user to a more suitable functional path). Although both the alternatives allow session adaptation in real time, the type of context considered is very limited. Considering a broader range of context sources is the research performed by Lungaro et al. (2010). It defines a predictive and context-aware multimedia content delivery for future cellular networks. Based on a Context Shift paradigm, it introduces software agents at the network side capable of monitoring both the users and the network. Then, based on predictive knowledge of mobile user behaviour related to consumption of content in smart devices, it delivers and stores content on the users' devices, before it is requested and consumed. This is done opportunistically by utilizing the instantaneous available excess of resources in the network. However, this strategy does not allow real-time streaming scenarios and minimizes the possibility to use context to improve group communications.

In next generation networks, multiple access networks will coexist on a common service platform. On the session layer, most of the solutions proposed use Session Initiation Protocol (SIP) as main signalling protocol on IP networks. It has been used in Multiprotocol Label Switching (MPLS)-based next generation networks (Know, 2006), addressing point-to-multipoint session management schemes and as an enabler for session mobility in converged networks (Oberle et al., 2007). Still based on SIP, the work in Pavlovski (2007) recommends IP Multimedia Subsystem (IMS) as key enabler for service delivery platforms. Some examples of such initiatives are the research works done by Bellavista et al. (2007, 2010). Although both of them aim at providing mobility solutions (i.e. handovers and session continuity), the first focuses on the terminal side, while the second on the network. The first enables dynamic determination and selection of the most suitable interface and connectivity provider among the available ones. For this, it considers several elements at very different abstraction layers (i.e. context types). Again, this limits its usage in group communication scenarios. The second approach deals with network-assisted handovers; however, it does not consider any type of context or intelligent stimulus to perform the aforementioned operation. For that, it will be necessary to have a set of network- and transport-aware mechanisms that will deal with such tasks.

Consequently, an access selection process, using any-constraint algorithm based on context, preferences and capabilities, should be in place to enable the optimization of both terminal and network (Jesus et al., 2008). Although many proposals base the decision process on radio signal properties (e.g. Pahlavan et al., 2000), this is only one of the many criteria in a network selection scheme. Thus, some proposals suggest that selection decisions should be context-aware (Jesus et al., 2008). Moreover, the majority of related work focuses entirely on network selection algorithms, not concerning other important mechanisms, crucial to support the decisions made by mobility protocols and QoS management, to enable the complete network re-configuration triggered by context. This lack of high-level perspective is addressed in more recent proposals (Chen & Yang, 2007).

In this sense, our approach goes towards an autonomic concept, by considering a dynamic optimization at the session, network and transport layers (the latter two with the help of external components). Furthermore, it considers the support of context-aware selection, in both unicast and multicast environments, where the group membership is kept flexible enough to support any parameter envisioned. Trying to cover the limitations identified in previous research initiatives, this work addresses the management of real-time multimedia sessions for both individual and group communications. Altogether, it pretends to achieve the optimal balance between user personalization and network efficiency.

7.3 System description

Foreseeing future personalized multimedia experiences, but at the same time, embracing current targeting needs for group communications, the system was designed to progress further and investigate the dynamic optimization of content delivery to a group of users based on its context – context-aware multiparty delivery. In this sense, the main objectives of this work are:

- Maximize user satisfaction in group communications.
- Address the heterogeneity of access networks, devices, environments and physical mobility by dynamic content adaptation.
- Achieve optimized context-aware multiparty delivery, by defining a novel, functional delivery architecture.

Given the above goals, optimizations are considered at several levels of the TCP/IP protocol stack, including session, network and transport layers. Furthermore, the context considered to drive the dynamic adaptation of the multiparty delivery encompasses not only the networking context of each group member (e.g. link quality and characteristics), but also their environmental (e.g. device capabilities, physical location, speed, etc.) and personal (e.g. preferred language, alone or with friends) contexts. Naturally, different adaptations can be applied, simultaneously, for different members of a group receiving the same content, with adaptations for each user being driven by its own networking, environmental and personal contexts.

7.3.1 The architecture

To better understand following concepts, this work needs to be contextualized into the framework developed under the C-CAST project. Figure 7.2 depicts a high-level architecture abstraction of the proposed framework, highlighting the entities where the multimedia delivery contribution is most notable.

Although not all of these entities are translated into the advertising framework (see Figure 4.9), there are some parallelisms that can be made. All the context-associated entities are reflected within the human enabler, the services and enablers are substituted by the advertising service, and the content features are minimally attached with the session management enabler (see section 7.4).

Before getting into the details of the Session Management Enabler (SME), it is important to understand its relationship with the network and transport layers, and how it interfaces with them. While Figure 7.3 depicts the context-aware multiparty delivery framework and associated context mechanisms, the following subsections highlight the main functionalities performed at each level.

7.3.2 The network layer

The network layer comprises a series of components that provide intelligent end-to-end path and network selection. This is achieved by considering user, network and environment contexts, allowing terminals to be always best connected and receive content over multiparty sessions while benefiting from satisfactory QoE. It takes advantage of diverse context information, obtained from multiple sources, and allows multimode terminals to be always 'best' connected. By making use of context, it successfully drives intelligent network optimization, be it in terms of communication path, terminal interfaces or access technologies' selection. Furthermore, it assumes complex, dynamic and heterogeneous scenarios, where network events (link failures, handovers and traffic conditions) take place randomly

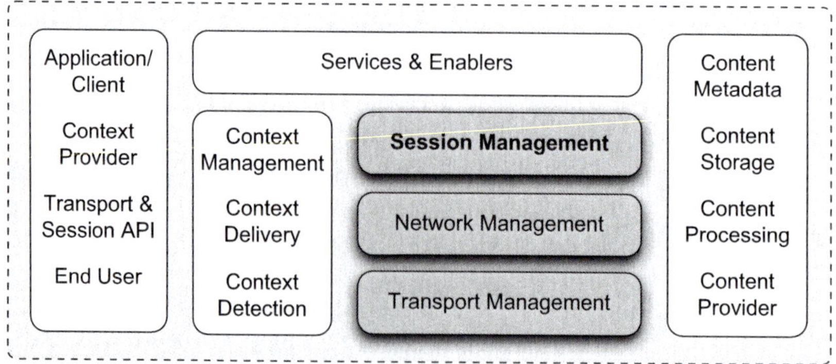

Figure 7.2 High-level representation of the C-CAST project reference architecture

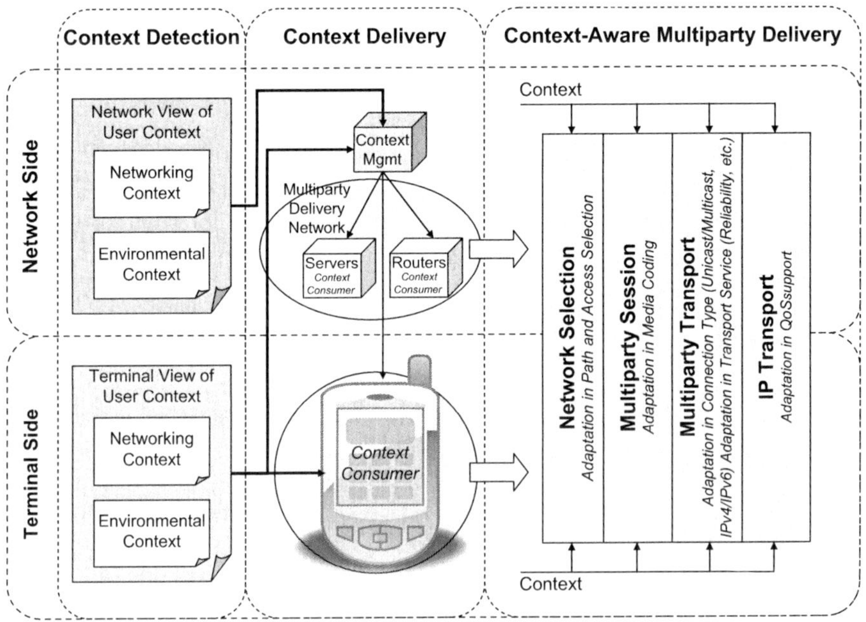

Figure 7.3 Context-aware multiparty delivery

over time. Such network dynamism and complexity require a new concept of network architecture in order to efficiently support management of users in sub-groups (process is explained in detail in section 7.4.3).

Providing interfaces towards upper layers, it has the responsibility of sorting out network-level groups. Actually, it divides the service groups into smaller sub-groups according to the quality on the stream they will be receiving. By taking into account current networking and environment user context (i.e. noise, interference, networks within reach, signal strength received, signal strength alteration rate, terminal's capabilities, user's preferences, etc.), as well as the current network context (i.e. QoS capabilities of Radio Access Technologies (RATs), available capacity and current load in RATs, etc.), it selects the RATs that will better serve each user and the content coding that each user will receive, in a way that enhanced capacity and network performance are achieved. The network selection process also provides support during the multiparty sessions' transport, that is users of ongoing sessions can receive multiparty content from a different network interface, due to changing network conditions or even vertical handovers (with the support of a Mobility Controller (MoC) installed in the terminal). By using standards when possible, its components map the QoS requirements of the multiparty session into an available class of service, also taking into account all the network status when selecting the path. In this sense, they deploy admission control operations along the network communication path. Afterwards, they finally take the decision of enforcing the reservation.

7.3.3 The transport layer

The main role of this layer is to enable Multiparty Transport Overlay (MTO). The goal behind using it is to provide a generic transport service for multiparty applications, that is supporting a variety of existing and future multiparty applications in a heterogeneous networking environment (e.g. WiFi, 3G, etc.) with changing networking and environmental contexts (e.g. network multicast capability, packet loss ratio, etc.). The MTO is a new concept in that it applies the overlay paradigm at the transport layer, while most of the existing multiparty overlay solutions, whether in the conceptual stage or deployed in today's networks, either use some form of IP tunnelling or specify application layer protocols (Moen, 2004).

In essence, the MTO should be generic (vis-a-vis the communication layers below and above the transport layer), scalable and reliable. Also, this overlay should provide a dynamic multiparty transport group management service, adaptable to the networking and environmental contexts. This context-aware transport group management paradigm is a key feature of the MTO. The transport group management per se represents a set of operations performed at the transport layer, on the whole multiparty group or on individual groups of users. This is a new concept, and no substantial work exists in the technical literature.[1] In the present work, this concept is closely related to the management of multiparty transport connections (i.e. creation, updation and deletion operations) according to the changing networking and environmental contexts. A multiparty transport connection is the set of all unicast and multicast connections of a given multiparty group. Conceptually, an MTO is a transport tree made up of Overlay Nodes (ONs). The root of the tree represents the multiparty source (e.g. Media Delivery Function (MDF) server). The leaf of the tree is the closest node to the receiver (e.g. user terminal). To properly deliver multiparty packets from the source to the receivers, over the overlay tree, each overlay node has to maintain mapping information between the multiparty transport connection ID[2] and the IDs of associated multicast and unicast transport connections, forming the branches of the MTO tree. Multiparty data routing (unicast or multicast routing) from the source to ON, between ONs, and from ON to receivers is handled by a specific module (also on the transport layer). Figure 7.4 shows an example of an MTO maintaining a multiparty transport connection made up of three multicast connections and three unicast connections.

Usually triggered by the network layer, MTO also provides an efficient transport framework able to optimize the use of network resources (e.g. by maximizing the use of IP multicast when possible) while adapting to the specific context of each group member (e.g. by using IP unicast where multicast is not available). In terms of context-awareness networking, the MTO component also

[1] Group management is a quite generic term, which was defined from different perspectives (e.g., group membership, mobility, media selection, group partitioning, etc).

[2] A transport connection is identified by <source address, source port, destination address, destination port>. Yet, a new ID type should be defined for a multiparty transport connection (e.g. a random number).

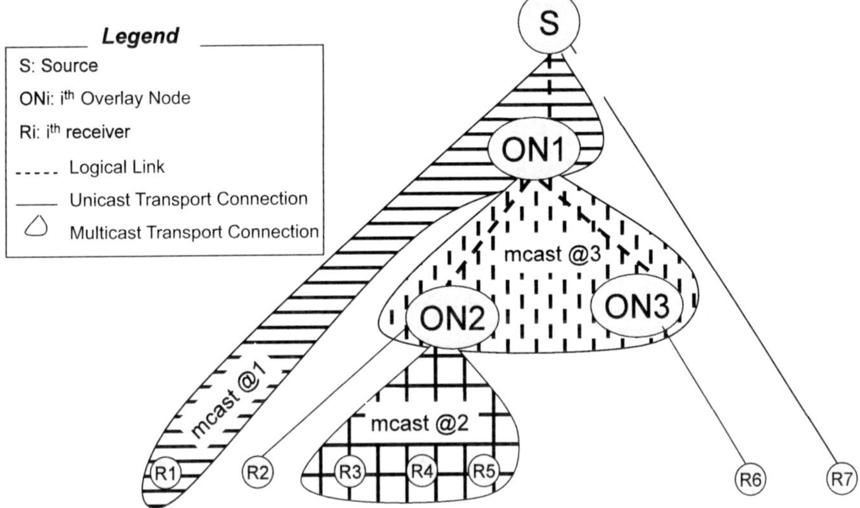

Figure 7.4 Concept of multiparty transport overlay

maintains adaptive transport reliability (e.g. adaptive Forward Error Correction (FEC)) based on the link quality.

7.3.4 Innovations

Altogether, combining the work done at the different layers allows: (a) making networking and environmental contexts available and (b) using this context information to drive the following adaptations for context-aware multiparty delivery:

- Adaptation in access network selection, including: (a) context-aware RAT selection for mobile terminals aiming at enhanced network capacity and performance, and (b) 802.21 and SIP-based terminal mobility between accesses triggered by network or terminal.
- Adaptation in routing path selection, consisting in selecting the best multiparty delivery path, based on network characteristics (bandwidth, load, delay, jitter, loss rate, multicast capabilities, etc.) and QoS needs from the session.
- Adaptation in media coding selection (called sub-grouping), consisting in efficient and scalable selection of best media coding per sub-groups of users, based on available content formats, device capabilities (e.g. codecs, resolutions supported), user preferences and network QoS capabilities.
- Adaptation in user device selection, with context-aware SIP-based user session mobility between terminals triggered by network or terminal.
- Adaptation in transport connection type, based on a MTO providing an efficient transport service for group communications, by hiding heterogeneity of underlying networks in terms of IP multicast capabilities or IPv4/v6 support. MTO tree configuration and update is driven by characteristics of multiparty delivery paths to receivers.

- Adaptation in transport reliability, through an adaptive FEC-based reliable transport service, well suited for streaming services with stringent latency constraints.
- Adaptation in QoS support, through dynamic QoS enforcement along the multiparty delivery tree (with unicast and multicast branches).

7.4 Session management enabler

The previous section introduced the global context where the SME can be inserted and the motivation of its usage within multiparty delivery for multimedia advertising environments. This section focuses on the particularities of the SME, how it interacts with other components, what are its functionalities and how it can be used in real case scenarios.

7.4.1 General overview

The session management enabler is responsible for managing all the user-to-content and content-to-user relationships. In fact, it provides the necessary signalling to deliver content to its consumers, handling different types of events, specifically, session initiation, modification, termination and mobility. Thus, it participates in dynamic changes, for example switching between different content for the same group of users because of new quality constraints. Since session management is closely interlinked with media delivery, it is responsible to ensure that content is delivered to its customers. This makes it the appropriate component to handle the coupling of context-awareness and broadcasting/multicasting technologies. Nevertheless, this is done with the support received from the network and transport layers. The delivery itself is performed by the Content Processor and Delivery (CtPD) component, which is responsible for streaming the content towards the users or to the ONs, when multiparty delivery is selected. Mainly due to the advertising industry requirements, the SME is capable of functioning in two modes. The first is called 'static' and assumes that a list of users, contexts and delivery conditions are pre-defined. This is particularly useful when the SME is used in conjunction with external components, which have their own reasoning algorithms or by entities that possess a list of users for a specific campaign. The second, named 'dynamic', allows advertisers to specify the user contexts they are interested, as well as the triggering options to deliver the advertisement. All the logic behind this is dealt at the Context/Rule-Based Engine (CRBE) module, implemented within the SME. To allow better understanding of how the SME works, the following subsections highlight its architecture, functionalities, interfaces and present some general use-cases. Furthermore, emphasis is given towards some design decisions taken under this work.

7.4.2 Architecture, interfaces and functionalities

This section will focus on the interfaces between the different modules that compose the SME, as well the ones enabling the communication with external components. To simplify the analysis of such reference points, Figure 7.5 depicts the functional representation of the SME and adjacent components.

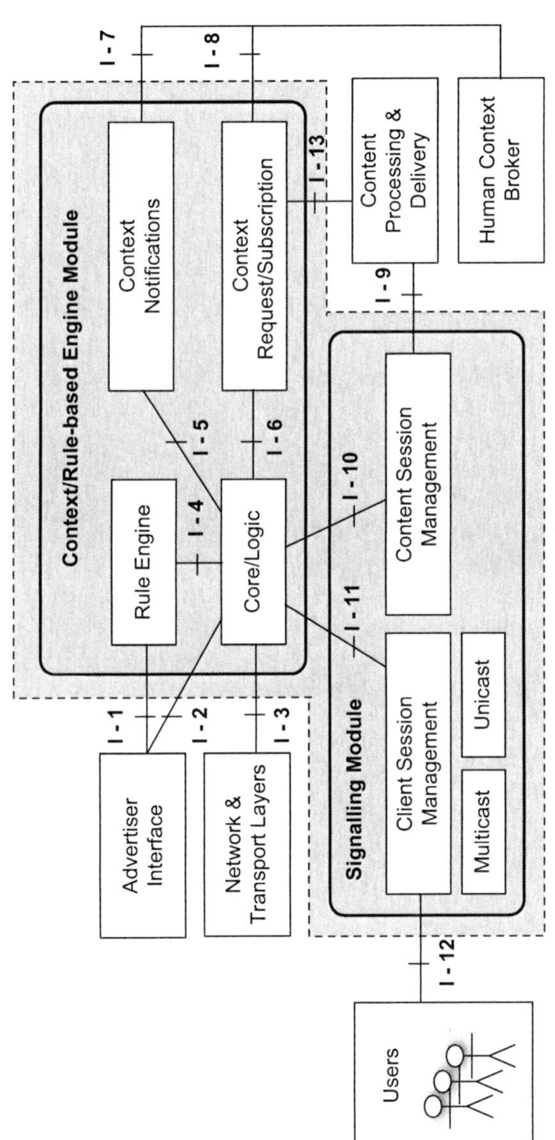

Figure 7.5 Session management enabler functional architecture

Before getting into the details of what happens at each interface, it is important to understand what are the functionalities of each of the sub-modules present at the signalling and context/rule-based engine modules.

7.4.2.1 Context/rule-based engine module

- *Core/logic*: responsible for managing and processing all requests coming from outside. Its interfaces allow direct access to recently cached information (context has always an expiration date) or an Application Programming Interface (API) for historical data. For the case of real-time context, it can be requested or subscribed. If the latter occurs, when context changes, a notification is sent to the previously subscribed entity. Furthermore, it allows information to be updated or created using a specific ContextML format.
- *Context request/subscription*: capable of requesting user-related context to the Human Context Broker (HCB), on behalf of the core sub-module. The request can have a synchronous or asynchronous nature. In synchronous mode, the reply is received inside the same entity, and forwarded to the core for further processing. In asynchronous mode, a notification URL is provided within the subscription parameters. This notification is processed further by the context notification sub-module.
- *Context notifications*: whenever a context change occurs, a notification is sent to the respective notification URL. This entity is responsible for receiving and parsing it, and subsequently forwarding the result to the rule engine, where further actions will take place (depending on the result).
- *Rule engine*: is an expert system that uses knowledge representation to facilitate the codification of knowledge into a knowledge base, which can be used for reasoning (i.e. we can process data with this knowledge base to infer conclusions). In other words, it is a system that uses rules, in any form, and that can be applied to data to produce outcomes. These data are usually passed through the context notification sub-module or from the core itself. The knowledge base/rules are set up using the advertiser interface, where campaigns are transformed into rules and actions.

7.4.2.2 Signalling module

- *Client session management*: triggered by the logic/core sub-module, it supports all the signalling process between the server and the client sides. Using SIP as signalling protocol, it enables a different set of scenarios, as the ones covered in section 7.5. The main reason why it is decoupled from its equivalent content sub-module relates to the fact that due to the nature of the proposed architecture, client-side changes will, most of the time, not affect content-side sessions, and vice-versa. Furthermore, it supports both unicast and multicast delivery methods. Nevertheless, there are some differences in the way this sub-module works for both scenarios (see section 7.5).
- *Content session management*: similar to the aforementioned client session component, it is responsible for managing the signalling processes with the

content processor and delivery. Basically, it controls the active sub-groups, by changing the session according to the instructions given by the logic/core sub-module. In the most common scenario, it is used to force the CtPD to stream to the Overlay Source Node (OSN), which is then distributed across the overlay tree until the Overlay Leaf Nodes (OLNs), the final nodes before reaching the users.

7.4.3 Design principles

Although the main concern when conceiving the SME was the fulfillment of the requirements specified in section 3.3.3, during its design phase, there were some principles that were considered:

- Minimize changes in the user terminal.
- Enable fast adaptation to a context change.
- Maximize reuse of existing standards when applicable.
- Distinguish between context-aware decision making and enforcement.
- Place intelligence in the network, where most of the networking and environmental contexts needed to drive adaptations in multiparty delivery are available.

To accomplish such ideologies, there were two main aspects that are worth highlighting: the *sub-grouping* and the *modular approach*. Despite they don't cover all the principles, they are eventually addressed under these two 'umbrellas'.

7.4.3.1 Context-based sub-grouping

Within the specified framework, a group is formed based on similar context conditions. In the case of an advertisement, this is represented by a set of context information specified by the advertiser (e.g. age, gender, location, time of day, etc.). Although this is extremely important for personalization and relevance, it might be very demanding from the network perspective, as the user's personal, environmental and network contexts can be different. In this sense, this work proposes a second grouping operation called sub-grouping. Its main purpose is to achieve a balance between personalization and efficiency, and occurs at the session and network layers.

The sub-grouping process starts once a new group/session is created by the application layer. Once this is done, the logic module receives the list of users, a content identifier and a session type. Based on these parameters, it executes three different but complementary actions (further details about the attributes and communication aspects can be found in section 7.5):

- Translate the content identifier into content descriptions for the available content streams (that the CtPD has stored or is capable of transcoding) – this is done through I-10.
- Request the context types the SME should request and subscribe, to perform sub-grouping and subsequent adaptations – this is done through I-4.
- Obtain user-related context, necessary to enforce initial sub-grouping – this is done through I-6 and I-8.

When all the information is gathered, the core sub-module is responsible for applying a filtering algorithm, where it matches the user context and capabilities with the content availability. In other words, for each user, it selects a set of possible media streams. The filtering parameters can be device oriented, like supported codec and screen resolution, or more generically, context-based, where preferred language, noise or movement can be used as input parameters to filter the selectable streams for each user. Once this is done, the filtered user information is sent to the network layer to decide which streams (audio, video and text) are more appropriate for each user. It is important to notice that the decision process could also be enforced at the SME (by requesting user network context information); nevertheless, when available, the components located at the network layer should be preferably used. Once the process is concluded, each user will be assigned with a combination of different streams (usually one audio, one video and possibly one text). Figure 7.6 depicts the whole mechanism.

Once the network layer notifies the SME, acknowledging that the resource reservation setup is concluded and delivering the list of users, respective sub-groups and media types, there is a need for the SME to parse this message, extracting the required information to initiate its Back-to-Back User Agent (B2BUA) functions (details are provided in the use-cases presented in section 7.5). As mentioned, although all users will be receiving the same content, the media types may vary according to their preferences, needs or current conditions, allowing each user to receive their own personalized set of streams. In this sense, a subset

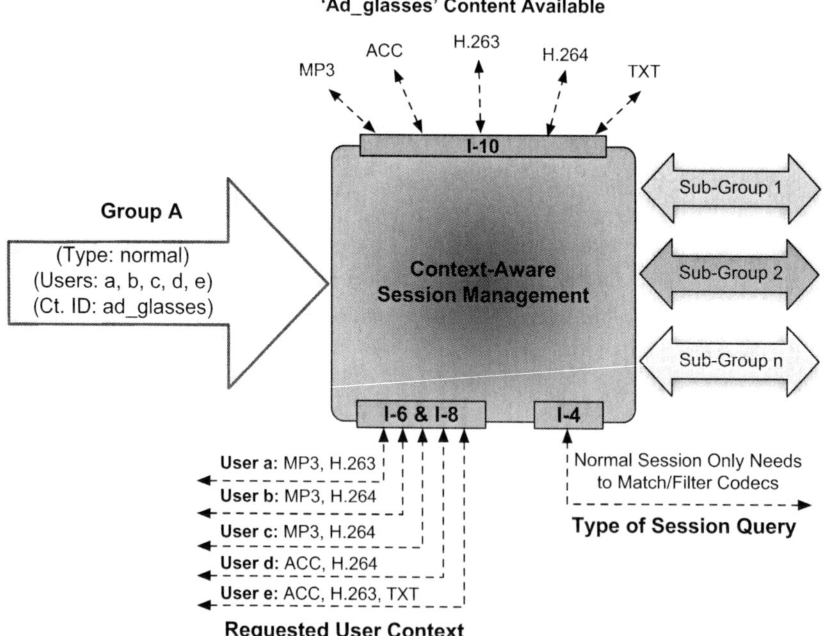

Figure 7.6 Logical representation of sub-grouping based on context

Audio	Video	Text
ACC	H.263	English
MP3	H.264	

(a)

	Audio	Video	Text
john@open-domain.org	MP3	H.263	English
maria@open-domain.org	MP3	H.264	---
lena@open-domain.org	ACC	H.264	English

(b)

Figure 7.7 Example of different media types per user for the same content

of the users might belong to the same sub-group concerning one media type and different sub-groups in others. The example depicted in Figure 7.7(a) represents all the sub-groups that were selected by the network layer to be created at the session layer. In this specific example, it is possible to identify two videos, two audios and one text sub-group.

Concerning Figure 7.7(b), it is easy to identify that user 'John' and 'Maria' are in the same audio sub-group but in different video sub-groups. Equally visible is the fact that 'John' and 'Lena' have the same video sub-group but all other do not match. At first sight it might seem illogical to create five sub-groups, when it would be sufficient to have three, with the specificity of each user. However, in reality, this system is designed to accommodate scenarios where big groups of people, with different requirements, are involved. In this situation, a one-to-one sub-group personalization would not make sense. Furthermore, by separating each media type into a different overlay tree, and consequently assigning a different multicast address, it is assured that the same media is leveraged just once until the users nearest node (OLN). In the other scenario, for two users receiving the same video stream (with different audio combinations), it would have to go twice throughout the entire overlay tree. Therefore, when several users are involved, it is very likely that the combination of all possible media types would be necessary. This can be reflected by the following equation:

$$S = (a + v + t) + (a \times t + a \times v + t \times v) + (a \times t \times v) \tag{7.1}$$

where S is the total number of sub-groups (and consequently overlay trees) that need to be created, a the number of possible audio codings, v the number of possible video codings and t the number of different text subtitles. On the other hand, the number of sub-groups required by the this approach equals the number of eligible media streams for each session $-a + v + t$. Covering most of the scenarios up to four variants per media type, Table 7.1 presents possible arrangements of different media types, while Figure 7.8 shows how the number of necessary sub-groups varies with the media types available/eligible for a specific content for both approaches (based on the formulas introduced earlier). Please notice that in terms of overlay networks or sub-grouping, there is no difference between two audio, two video, one text and one audio, two video and two text media types.

Although Figure 7.8 depicts a very heterogeneous situation (assumes all possible combinations exist), in reality, due to different devices, network and environmental context conditions, it is very likely that the real numbers tend to this scenario.

Table 7.1 Possible scenarios of media types available

Scenario	A	B	C	D	E	F	G	H	I	J	K	L	M	N	O	P	Q	R
Audio	1	1	1	2	2	2	3	3	3	3	3	4	4	4	4	4	4	4
Video	0	1	1	1	2	2	1	2	2	3	3	1	2	2	3	3	4	4
Text	0	0	1	1	1	2	1	1	2	2	3	1	1	2	2	3	3	4

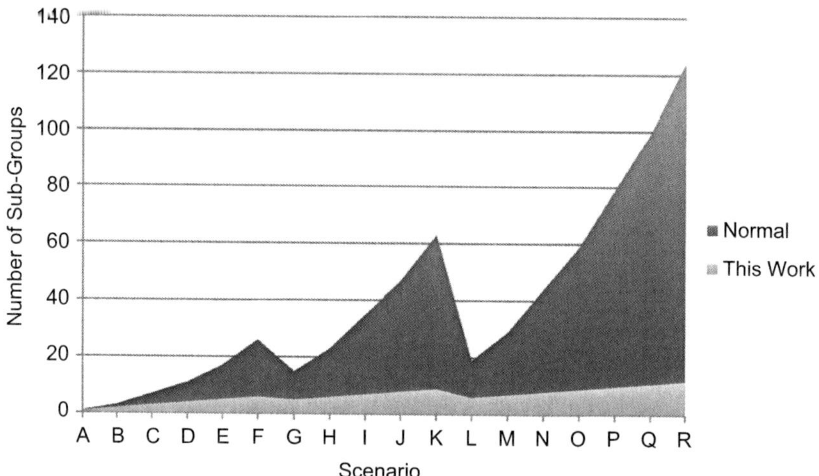

Figure 7.8 Number of necessary overlay trees for approach with and without sub-grouping

However, despite this methodology may simplify the configuration and increase efficiency on the network side, it is not without disadvantages. As the client will now receive more than one media stream (one for each media type), it must assure that all of them are synchronized. This can be done by adjusting the jitter buffer dynamically. Nevertheless, as it will be shown in section 9, under the evaluation, this situation will not be a major issue. Furthermore, although outside the scope of this work, such approach (one media type = one sub-group = one overlay tree) will enable future clients to dynamically adapt the way multi-media content is consumed, without direct intervention of the network side.

7.4.3.2 Modular approach

In theory, there are several reasons lead to the adoption of a modular approach. These include improving efficiency, reliability and scalability. In practice, in this framework, it can be seen at different levels to:

- Differentiate application logic/core, from signalling and multimedia delivery.
- Separate application decision making from enforcement.
- Delegate different functions across the application, session, transport and session layers.

At first sight, these topics might seem unrelated, nevertheless they aren't. The application logic is usually coupled to the application layer, where the decisions are taken based on workflows or process and rules stored in the knowledge database. Signalling is then used to enforce decisions, on both the client and the server (where the media is hosted) sides. As the SIP or HTTP protocols are used to perform such actions, this is done at the session layer. Multimedia distribution is therefore the outcome of the aforementioned procedures and commonly benefits from previous network and transport, resource reservation and enforcement (also done previously).

Despite being modular, the logic lies on the capabilities of a single sub-module, the rule engine. With it, the framework inherits the following advantages (http://www.jboss.org/drools):

- *Declarative programming*: allows the user to say 'what to do' and not to worry with 'how to do it'. The key advantage of this point is that using rules, one can make it easy to express solutions to difficult problems and consequently have those solutions verified. Rules are much easier to read than code. Furthermore, they provide an explanation of how the solution was achieved and why each 'decision' along the way was made.
- *Logic and data separation*: the data is gathered from the interfaces with the HCB and transformed into domain objects, while the logic is in the rules. The upshot is that the logic can be much easier to maintain when further changes are required in the future, as the logic is all laid out in rules. This is especially true if the logic is cross-domain or multi-domain (different contexts or different advertisers). Instead of the logic being spread across many domain objects or controllers, it can all be organized in one or more very distinct rules files (e.g. campaigns or basic system rules).
- *Speed and scalability*: being a forward engine system, and using specific algorithms, it provides very efficient ways of matching rule patterns to a specific domain object data. These are especially efficient when datasets change in small portions, as the rule engine can remember past matches. These algorithms are battle proven.
- *Centralization of knowledge*: by using rules, it creates a repository of knowledge (a knowledge base), which is executable. This means it is a single point of truth for business policy. Ideally rules are readable, and therefore, they can also serve as documentation. These rules are the outcome of a campaign definition, set up at the advertiser interface.

To allow understanding of how a forward chaining rule engine works, Figure 7.9 depicts the flow of operations between the different stages. Basically, this model is 'data-driven' and thus reactionary, with facts being asserted into working memory, which results in one or more rules being concurrently true and scheduled for execution by the Agenda. In short, it starts with a fact, it propagates and it ends in a conclusion. In principle, when a 'positive' conclusion is achieved, it results in some sort of multimedia (advertising) delivery.

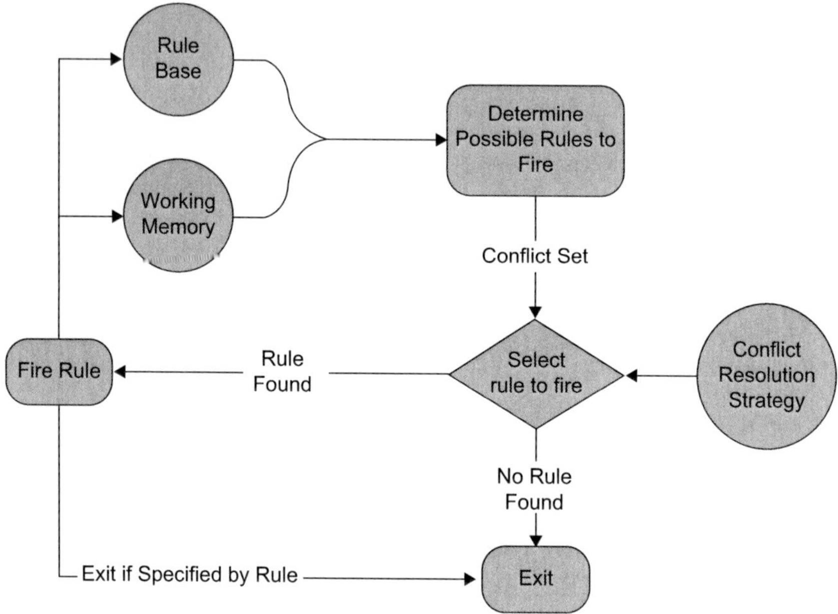

Figure 7.9 How a Forward chaining rule engine works

The modularity can also be seen from a layered perspective, as mentioned in section 7.3. Despite this component deals with application and session layers exclusively, it provides the necessary interfaces towards the network and transport planes.

7.5 Use-cases

To enable a better comprehension of how the system works, namely multimedia delivery, this section presents a set of use-cases. These will highlight some of the features and capabilities initially identified as requirements in Chapter 3. Furthermore, each use-case will be explained with a practical example within quotidian or futuristic advertising scenarios. Note that for same use-cases, depending on the user contexts available, or components involved (e.g. external components working at the session and network layers), the flow of information may slightly vary. To avoid possible misunderstandings, the examples provided will be as complete as possible, although simple scenarios may be used for different situations. Table 7.2 contains the description of the acronyms used in the use-cases' sequence diagrams, as well as a reference to where more details about the components can be found.

Table 7.2 List of acronyms to be used in the use-cases sequence diagrams

Acronyms	Description	Reference
AI	Advertiser interface	Section 4.3.1, page 71
CRBE	Context/rule-based engine	Section 7.4.2, page 140
CtPD	Content processor and delivery	Section 4.3.1, page 71
HCB	Human context broker	Section 4.3.1, page 71
NTL	Network and transport layers	Section 7.3.1, page 135
SM	Signalling module	Section 7.4.2, page 141

7.5.1 Session initiation – create a session with several users

7.5.1.1 Scenario description: at a sports event

This scenario illustrates a popular scene in our daily lives. Imagine a typical sports event where users from different teams are physically in the same area, connected to different networks, using different devices and obviously having different preferences. Usually, during the break time, users are more likely to permit advertising. Therefore, it might make sense to set up time as the advertisement trigger. Location can be used to select the users eligible in this offer (e.g. when they are in the vicinity of the bar), while preferences may allow the system to perceive which club each user is supporting. Alternatively, this information could be extracted from the users' social networks. Together, the network conditions and the device type may perform a critical role when deciding (it is the system who decides) the multimedia type (e.g. codec, resolution, frame rate, etc.) in which the advertisement will be delivered. The content itself can also vary due to weather conditions, number of friends together in this event, location in the stadium (e.g. VIP vs. common seats), etc. To sum up, different multimedia sessions may be created to better target the users' expectations. Nevertheless, this scenario will focus on a single session (i.e. same content for all participants) with several users involved.

7.5.1.2 How it works?

The best way to understand how the session initiation process works is by looking at a message sequence chart between all the intervenient components. Figure 7.10 represents these flows. As mentioned before, the scenario considers a complete setup where the network and transport layers (NTL) are also involved. However, the details of what happens there are outside the scope of this work.

By going through Figure 7.10 it is possible to understand that the first steps (1–5) try to select the list of users eligible for the advertising. Whenever an advertising campaign is set up (1), the CRBE subscribes (2) to all the contexts types necessary to evaluate the conditions and rules contained in the previous request. Before notification event triggers the delivery, it is likely that the pre-selected set of users change their context. When this occurs (3), the HCB is updated and all the entities subscribed to this context type are immediately notified (4). Afterwards, this

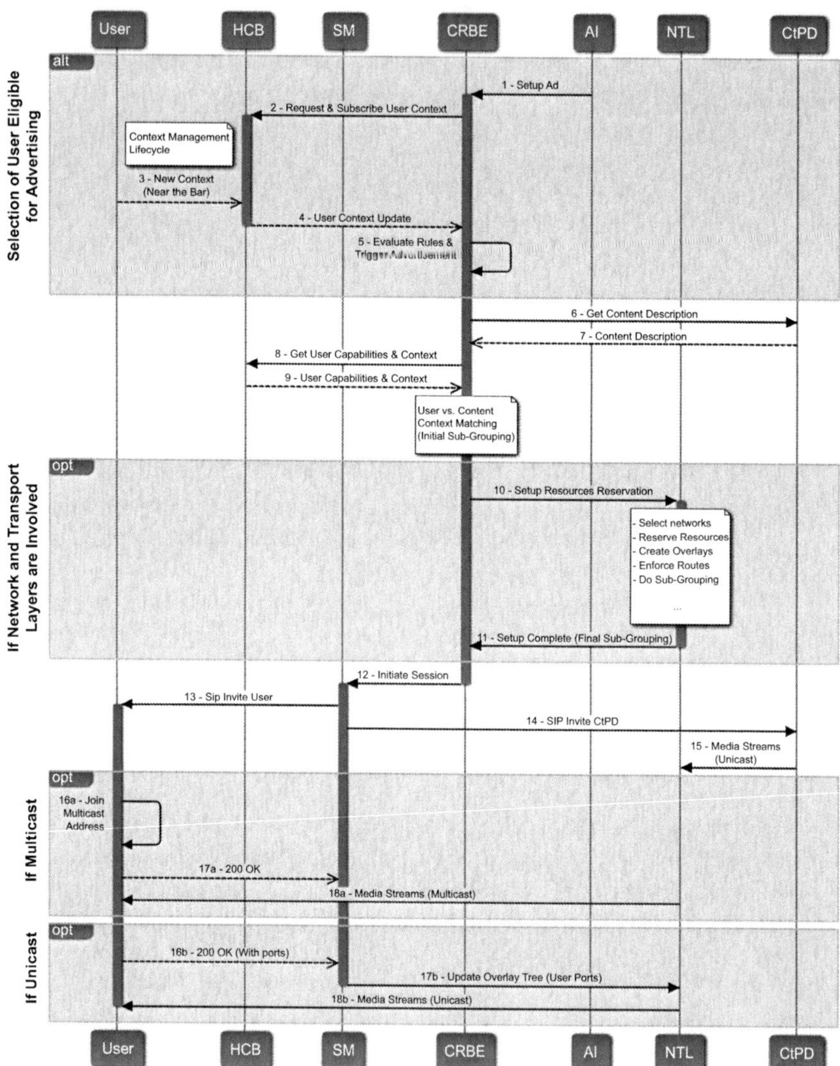

Figure 7.10 Overview of the session initiation message sequence chart

notification can or cannot trigger the advertising delivery (depending on the initial setup). In this example, it does not. Only when a specific time constraint (i.e. the event break occurs) is met, the triggering procedure occurs (5).

At this step the CRBE is aware of the list of users to include in a session and the multimedia content to be sent to them. Therefore, the first step is to get more information about the type of multimedia available for the specific content (6). Once the response is received (7), the CRBE knows which media types the CtPD has stored or is capable of reproducing (i.e. using transcoding techniques). A similar query is done towards the HCB, where a list of users is sent (8), with the purpose of retrieving (9) their current contexts and more concretely, their capabilities (e.g. supported codecs, screen resolution, etc.). After these phases, the CRBE is in the possession of all the information from both clients and content.

Based on this, it performs a first matching, filtering for each user, which content types it is capable of receiving under the current conditions. When the NTL are not available, the sub-grouping is performed at this level according to the system settings (e.g. choose the most efficient codecs or the highest quality supported). Nevertheless, when available, a resource reservation setup message is sent (10) towards the NTL. This message includes the list of users and the respective filtered supported content types. With these data, and according to the operator policies, the best network is chosen for each user (i.e. when the device supports more than one). In parallel, a set of overlay trees are built (one for sub-group), paths are enforced, resources are reserved and users are allocated to the different sub-groups. This is the final step of sub-grouping. Detailed information about this process can be found in section 7.4.3 and in the work developed under the C-CAST project, namely Simoes et al. (2009) and Antoniou et al. (2010). When all the network and transport operations are finished, the CRBE receives back the list of users, their addresses (i.e. IP and ports), settings and the associated sub-groups (11). This message triggers the communication with the signalling module (12). Depending on the implementation, the Signalling Module (SM) can invite the user (13) or the CtPD (14) in the first place. Usually this is done in parallel or after some threshold value. When receiving this message, the CtPD is responsible for establishing a unicast session with the OSN for that specific sub-group (15). This process is repeated for each sub-group, whose media details are contained in the invite message (14). It is then the responsibility of the NTL to assure the correct transportation of the media until the OLN, the nearest node to the user. More details about this topic can be found in sections 7.3.2 and 7.3.3 and in Janneteau et al. (2009).

On the client side, the behaviour might change slightly, depending on whether the user (i.e. its device and network) supports multicast or only unicast. If the first scenario occurs, the user sends a join message to the multicast address (16a). Afterwards it signals the success of this operation with a 200 OK SIP message towards the SM (17a). At this point, it is already capable of receiving the multimedia streams (18a). On the other hand, when unicast is the only option, the user client selects the internal ports for the communication and reports them inside the 200 OK message (16b). When receiving them, the SM contacts the NTL (17b), informing that the overlay tree leaf node responsible for that particular user should start streaming to the addresses provided on the user response. When this change occurs, the user automatically starts to receive the appropriate multimedia streams (18b).

7.5.2 Session modification – new user joins current session

7.5.2.1 Scenario description: the shopping mall

It is very common that people attend shopping malls without a concrete idea of what they are buying. What if there was a service that could inform them about the latest promotions on their favourite brands, as soon as they walk in the shopping mall? It might not make sense to initiate a session for each user that enters the mall. However, it is likely that different user profiles are created and a session is initiated for each of them. Therefore, each shop can decide on which profile group to advertise. Assuming Elena (married, 46 years old) enters the mall and her profile preferences (or past profiled activities) indicate that she is an avid jewellery buyer. At the same time, the system identifies an on-going advertising multimedia session for this profile group. Therefore, she will be invited to join the same advertising experience. For each session, it is considered that a set of different advertisements are played sequentially and on a loop.

7.5.2.2 How it works?

Similarly to the aforementioned use-case, the interactions are more easily understood with a sequence diagram. Therefore, Figure 7.11 exemplifies how the different entities interact. This scenario assumes the user device and network are multicast capable and therefore the unicast scenario will not be covered. Nevertheless, the principles are the same as in the session initiation.

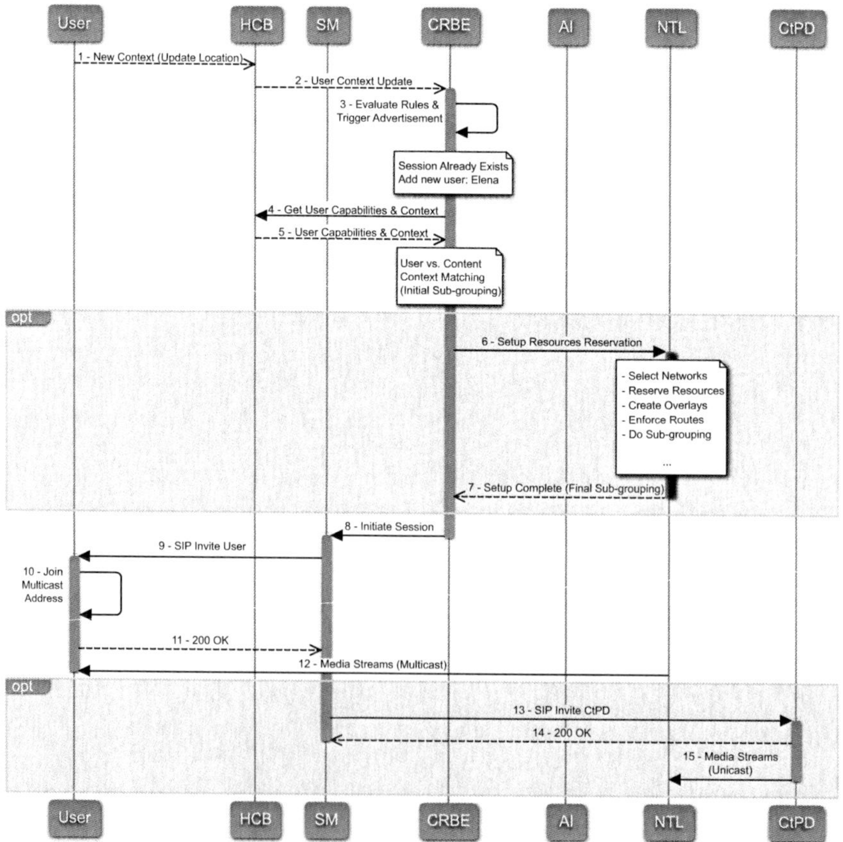

Figure 7.11 Overview of the session modification message sequence chart

When the user enters the mall, its location is updated in the HCB (1). Consequently, it notifies all the subscribed parties with the new location of the user. Based on the user context and the pre-defined advertising campaign requirements, the system decides (3) that Elena should receive a specific type of advertisement (i.e. based on the profiles mentioned before). Assuming that other users are already experiencing this specific content (i.e. session was previously initiated), the user Elena will be added to the same session. To find out what are the user capabilities, the CRBE requests the user current context and capabilities (4). Once these are received (5), the CRBE contains all the necessary information to perform the initial filtering process. Please note that the content descriptions are already known because the session was previously initiated.

The following steps (6 and 7) are very similar to what happens in the session initiation. Depending on the sub-groups previously created, two different things can happen. On the first option, for this specific user, the NTL select a set of sub-groups already existing. In this case, it only needs to update its resources to respect the operator QoS policies. The second option presumes that the NTL select at least one sub-group that was not previously created. Independently from the situation, the user receives an SIP invite message (9) to join the multimedia sub-groups assigned by the NTL. Then, it joins the multicast addresses (10) and signals the success of this operation to the SM (11). After these steps, the user starts receiving multimedia (12). If there is a need to set up a new sub-group, the SM sends an SIP invite towards the CtPD (13), which is answered with a 200 OK answer (14) if the process is completed without errors. Once this happens, the CtPD starts a unicast session towards the recently created overlay tree (15) (within the network and transport layers). In this case, it is assumed that the CtPD has the capability to start streaming in the middle of a specific media. Otherwise, it is assumed that this sub-group was created at the session setup, but until this point, had no user attached to it (and consequently the packets were discarded at the CtPD or the OSN).

7.5.3 Session termination – user leaving, no more content to transmit

7.5.3.1 Scenario description: leaving the park

There are several reasons why someone might want to end a current multimedia session. The end user may no longer be interested in the content he/she is receiving or may have some other desire. Nevertheless, it can happen that the system itself decides that the current targeting is no longer appropriate, or its capacities are reaching its limits, and therefore some measures need to be taken. Alternatively, the situation where there is no more content to transmit may also occur. This scenario assumes the situation where an amusement park will be closing soon, and consequently, makes no longer sense to advertise to its customers. The triggering of this action may occur, because the users are not anymore in the proximity of the park, or due to a previously defined time setting. In the first case, the multimedia session remains active, while some users (e.g. the ones whose context no longer matches the targeting conditions) are excluded. In the second case, the entire session is terminated and all users are disconnected from the ongoing multimedia session. Explanation in the following sections will cover both alternatives.

7.5.3.2 How it works?

With the exception of the triggering methods, independently from the scenario, the system behaviour is very similar. It can be explained by the diagram in Figure 7.12.

Starting with the scenario where the user voluntarily decides to leave a specific advertising campaign, a message is sent towards the SM (1a). As this entity is only capable of managing the signalling (SIP) part, it needs to inform the CRBE about this intention (2a). In the second example, the HCB receives a context update (1b),

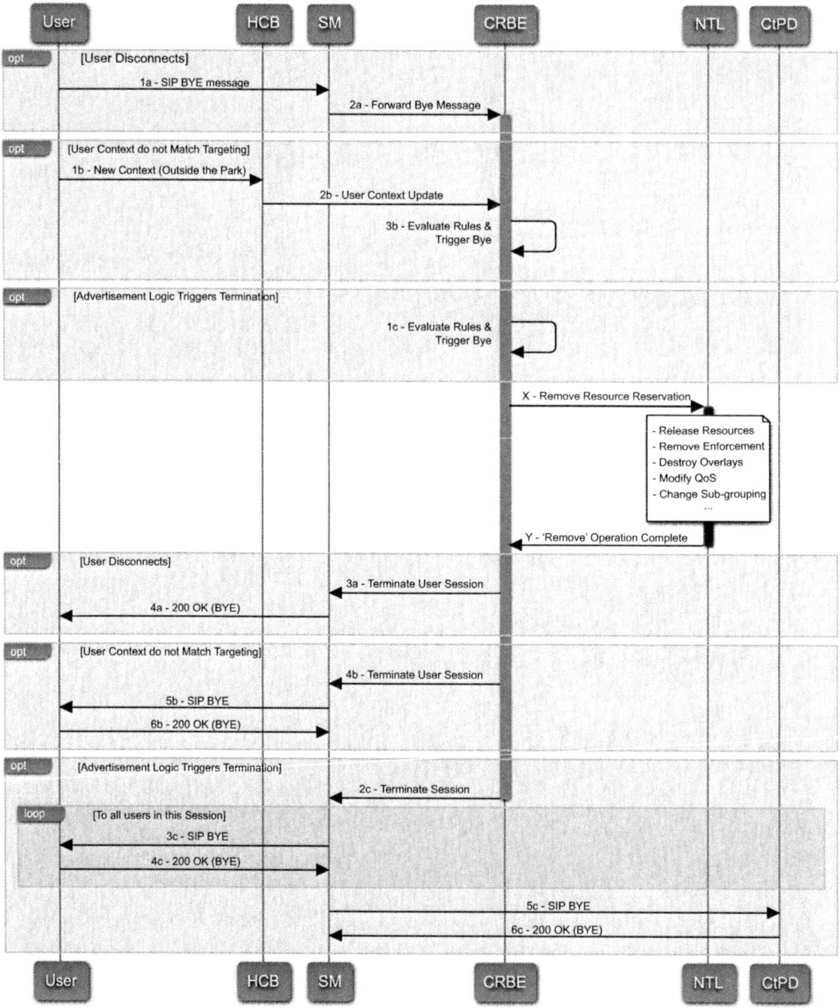

Figure 7.12 Overview of the session termination message sequence chart

stating that the user is no longer within the initially defined campaign radius. As always, the HCB sends a notification (2b) to the CRBE (which has previously subscribed for this particular context). When receiving a notification, the logic is triggered and the rules evaluated (3b). The outcome of this process states that the user should be removed from the current advertising session. The last scenario happens when the advertisement logic triggers a new rule evaluation (1c) and the outcome determines that the current time does not meet the pre-defined requirements (i.e. no advertising should be done 30 minutes before the park is closed).

After these steps, all scenarios are in the same situation. Therefore, the next action involves the communication with the NTL (X), where the previous resource reservation needs to be cancelled. Depending on the number of users to be removed, the actions performed on these layers may vary. When the user removal does not affect any sub-group, the system simply releases the resources reserved for this user. On the other hand, if the user removal affects any sub-group (e.g. was the last user on a specific sub-group), the overlay trees need to be destroyed and the policy enforcements to be changed. When the complete session is to be terminated, all the overlay trees and respective resources are removed. When this process is complete, the NTL notifies the CRBE of all the changes he needs to perform (Y). Nevertheless, the changes are always reflected in terms of signalling, and therefore, this notification is transformed into a request towards the SM (3a, 4b, 2c). For example where the user showed an intention to disconnect, the SM simple sends a 200 OK message (4a), confirming the will to leave the session. For the second case, the SM entity sends a SIP BYE message to the user (5b), which answers back with a 200 OK, acknowledging the decision (6b). For the last scenario, the SM sends a SIP BYE message to all the users involved in this session (3c) and waits for their acknowledgements (4c). In parallel, it terminates the existing session with the CtPD (5c) and waits for its answer (6c).

7.5.4 Context change – user enters in a noisy environment

7.5.4.1 Scenario description: arriving the music event

Especially over the summer, hundreds of music festivals dominate the cultural agenda all over the world. The majority of these events are mainly sponsored by brands, which most of the time do promote their services inside the event premises. This brings a great opportunity to advertise inside and near the event premises. To avoid big agglomerates, parking slots or public transportation is usually situated a bit far from the events (sometimes more than 1 km). This scenario explores the possibility to advertise from the point a user gets closer to the premises, until he/she gets inside. It is very likely that in places not so crowded, the user is targeted with a full multimedia experience, where both audio and video media types are present. Nevertheless, as the user gets closer to the entrance the environment noise gets higher. To maximize the user experience, when a specific noise threshold is achieved, the system is capable of ‘transforming’ audio (when it becomes imperceptible) into subtitles. This will enable the user to keep following the advertisement in favourable conditions. This trigger can occur due to a sensor situated on the user's device or some other environment sensor (depending on the application). Similar adaptations could occur, where the same principles would apply (e.g. while the user is moving the video is suspended to save bandwidth).

7.5.4.2 How it works?

From a general point of view, the adaptation techniques can be used to increase both the user's personalized multimedia experience and the network efficiency. It is up to the system administrator to make sure that a balance between both can be achieved. Therefore, when setting up a campaign, it is important to also think about the impact it may have on the operator network. To allow understanding of how the adaptation is performed, Figure 7.13 highlights the communications between the entities involved. Please note that there are several techniques that can be used to adapt a multimedia stream with audio and video to a video with subtitles without audio. To simplify the flow, in here, it is assumed the use of a different video file with subtitles embedded at the source (server side). Alternatively, the subtitles could be sent (there are several ways of doing this) to the user and rendered as an overlay on the client side.

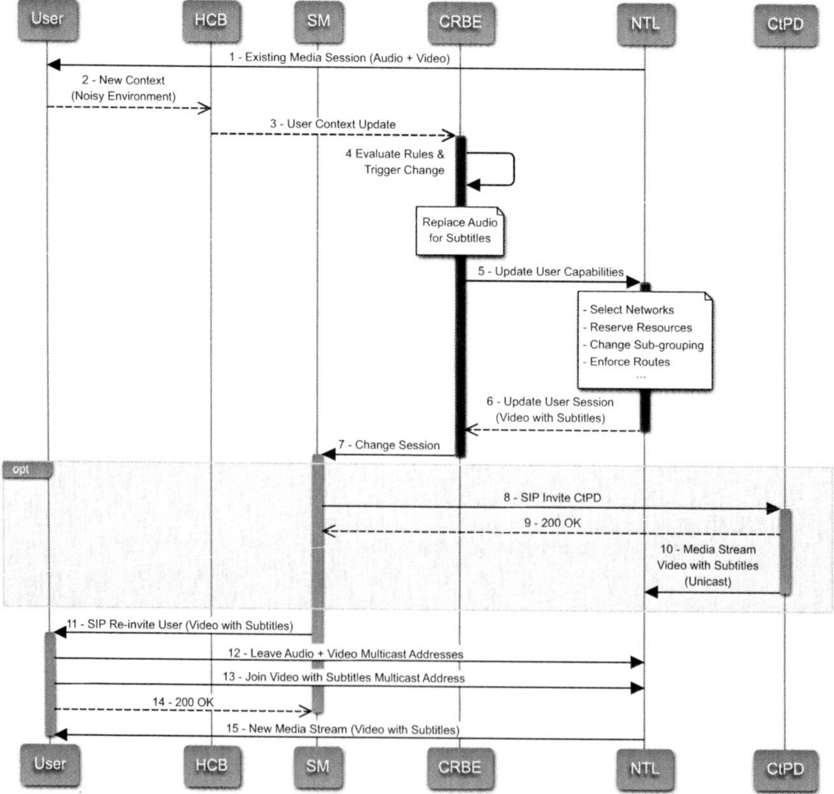

Figure 7.13 Overview of a user-context change message sequence chart

For this example, it is assumed the user is already receiving an advertisement containing audio and video (1). As before, whenever a context change occurs (2), the HCB forwards it to the subscribed parties (3). When it reaches the CRBE, the new context is evaluated (4) and according to the rules, whenever a noise environment is detected, the audio part of a multimedia session should be replaced by subtitles. Therefore, a request is sent to the NTL to update the session according to the new user capabilities (5). In fact, the capabilities are the same, nevertheless the CRBE removes them from the request (i.e. applies an additional filter). This way, the NTL are forced to find or create an appropriate sub-group where the video is encoded with subtitles. Due to the current network conditions, or user device limitations, it may happen that the new sub-group forces the device to change network (see section 7.5.5), or change the streaming quality. Either way, the network is reorganized accordingly and the new sub-grouping information is included in the response sent to the CRBE (6).

Afterwards, the adaptation information is forwarded towards the SM (7), which is responsible for the signalling process. If for some reason the sub-group was not previously created, the SM interacts with the CtPD to perform the necessary changes (8–10). After this step, the network is ready to provide the user with the new content type. Therefore, the user receives a SIP re-INVITE request (similar to an INVITE request but with new media descriptions) containing information about the video with subtitles' multicast group (11). When processing this request, the client leaves (12) the previous multicast addresses (audio + video), and joins the new video with subtitles' multicast address (13). After this action is complete, the user acknowledges this fact (14) and starts experiencing the context-aware adapted multimedia session (15). If in the future, the noise value for this specific user goes below the threshold, the inverse procedure may occur.

7.5.5 Handovers – session mobility vs. terminal mobility

7.5.5.1 Scenario description: daily home-work, work-home routines

For the majority of the adult population, the routine 'home-work-home' is done on a daily basis. With the development of the networks and devices, multimedia consumption on the move has become a reality. Together with it, advertising is not an exception. This scenario can therefore be divided in two parts.

The first illustrates the situation where John is using public transportation while travelling to work. As his trip is short, he likes to have a compact of the main news within his fields of interest. Due to monetary limitations, he prefers to have this service sponsored by advertising (for free). Still, he has to pay the fees associated with the 3G connection on his device. Luckily for him, the train company offers a free wireless connection to its customers. Therefore, at least while he is on the train, he can use this connection, which allows him to save money and experience the multimedia stream in higher quality (usually the throughput on WiFi networks is higher than in 3G). To enable such adaptation, the system needs to support terminal handovers.

The second scenario covers the opposite commuting path (work-home). Assuming John usually works until late, he cannot get home in time to watch his favourite team playing (the game is sponsored by his favourite brands). Therefore, he starts watching the game on his way home, using his mobile device. Although the quality of the stream and dimensions of the screen are not ideal for the type of event, it is better than missing the first minutes. When arrives home, John sees his multimedia session (the game with respective advertisements) automatically transferred to his high-definition television set. This handover type is called session mobility.

7.5.5.2 How it works?

The mobility aspects are adapted from the work done on the C-CAST project (http://www.ict-ccast.eu/). Consequently, some features are dependent on the NTL and associated client components. The previous work is particularly important for terminal mobility, as it requires some changes on a typical client device. For the session mobility scenario, the NTL are not mandatory, although important to assure QoS and other control mechanisms. To allow a better understanding, these scenarios will be analysed separately.

7.5.5.3 Terminal mobility

Terminal mobility can be triggered by the network or by the terminal itself (with the user consent or based on policies). Still, independently of the stimulus provision, the

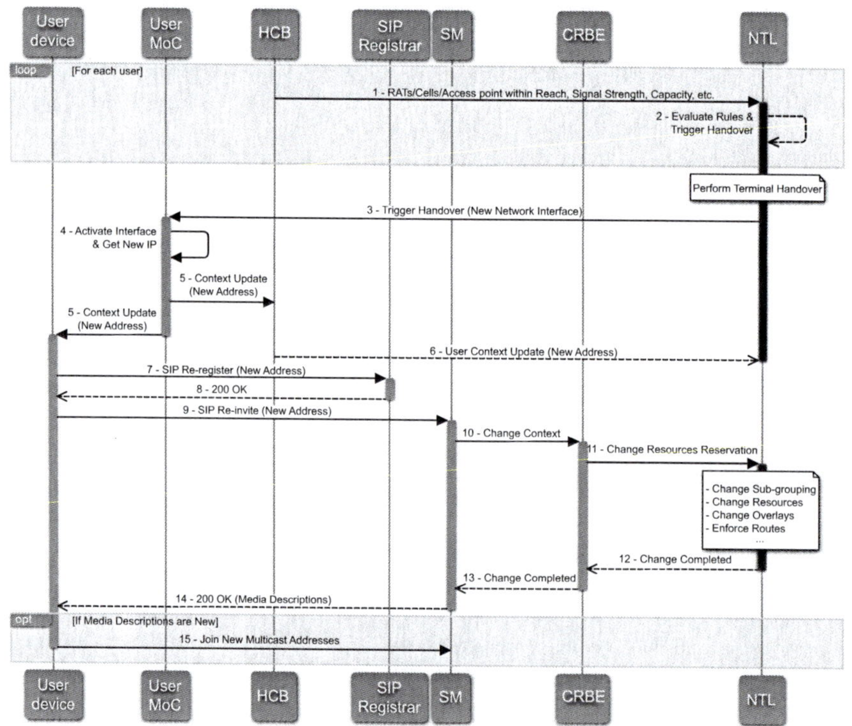

Figure 7.14 Overview of a terminal mobility message sequence chart

process is assisted by the network. The NTL take the decision of selecting the best RAT from the candidates provided by the CxP and triggers the handover process at the MoC, a module implemented on the terminal side. Figure 7.14 illustrates the whole scenario.

In order to address any change in the context and adapt to the most efficient transmission arrangement through time, the NTL periodically (period T) receive from the HCB the users' and network's instantaneous context (1). Based on the new context information received, the NTL execute the RAT selection algorithm and check (2) if a more efficient transmission arrangement than the one currently used exists (i.e. checks if the handover of some users between the RATs enhances capacity and performance). If a new transmission arrangement is selected, for each user that a new RAT/Cell/Access point is selected, the NTL trigger the handover of the terminal to the new RAT/Cell/Access point, by sending the new interface selected to the MoC, running in the terminal (3). Then, the MoC re-configures the mobile terminal to use the newly selected interface; more precisely, it activates the new interface and determines the IP address on that interface (4). After the traditional handover mechanisms, the terminal updates its addresses inside the HCB (5). As the NTL had previously registered to receive the user's IP address whenever a change occurs, the update results in a notification (6) towards the NTL. In parallel, the MoC triggers (5) the SIP UA on the terminal (and passes the new IP address) so that the latter initiates the SIP re-registration procedure with the registrar server (7), which promptly acknowledges this change (8). A similar procedure occurs towards the SM, where the SIP UA sends a re-INVITE message (9) containing the new address details.

Like in any other change (that involves multimedia), the SM signals the CRBE (10) to request (11) the necessary adaptations at the NTL. For this example it is assumed that the sub-grouping is not affected (the procedure would be the same as seen in the previous examples). Nevertheless, the NTL may need to update the overlay trees and respective resource reservation. After this point, the media should automatically be streamed towards the new user's address. After this step, assuming no changes occurred in the sub-grouping, the only remaining messages are simple acknowledgements throughout all the entities (12–14). When the new settings are known and the client/network is multicast capable, the client sends a JOIN (15) request to the new multicast address, contained in the 200 OK message (assuming the sub-grouping changes for this user). After this step, the user starts receiving the newly adapted multimedia content. If, on the other hand, the user only supports unicast, no changes are required as the adaptations are done at the overlay tree (more concretely, at the OLN).

Note that for a multicast-enabled device/network, it is possible do reduce the handover delay. For this to happen, the device might automatically join the previous multicast address, until it receives the 200 OK message (14) containing the new session descriptions. Nevertheless, its characteristics might not be indicated for the new network configuration. Another issue worth mentioning is that this type of handover presupposes a small service interruption and depending on the scenario, it may not be performed in a seamless way. This is one of the limitations of the task being performed at the session layer, where its simplicity contrasts with its limitations. To improve the quality of the handover, the lower layers must be involved in this process.

7.5.5.4 Session mobility

The other type of handover is called session mobility, and again, it can be triggered by the network or by the terminal itself (with the user consent or based on policies). Furthermore, independently of the stimulus provision, the process is always assisted by the network. In a few words, this process consists in transferring an existing multimedia session from a device to another (belonging to the same user). To be inline with the spirit of this work, the triggering being considered will have network origin, resulting from a user-context update. Figure 7.15 demonstrates this process.

Getting back to the original scenario, when the user arrives home, his/her device updates his/her location automatically on the HCB (1). This originates a notification from the HCB towards the CRBE (2), which triggers the rule evaluation mechanisms, resulting in a handover decision (3). This happens because somewhere in the policies associated with user John, there was a rule stating that all multimedia sessions should be transferred to his high-definition television set

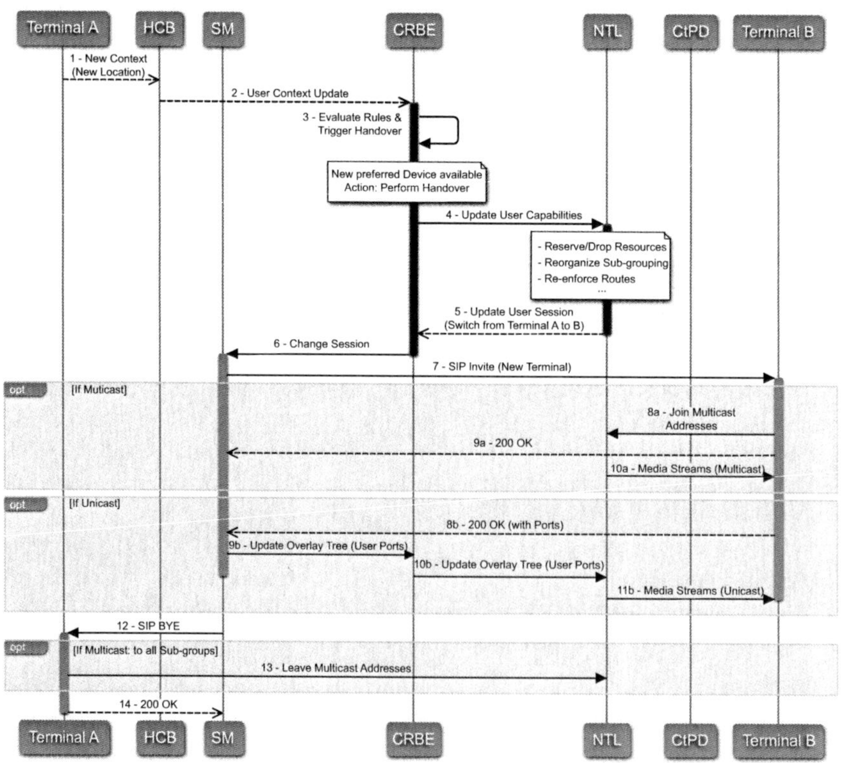

Figure 7.15 Overview of a terminal mobility message sequence chart

whenever his location was 'at home' (preferred location). Similarly to other changes, the CRBE requests the NTL to perform the necessary adaptations. Basically, resource reservations are changed from some sub-groups to another. To simplify this diagram, it is assumed that the new multimedia sub-groups associated with the new terminal (B) were previously created when the handover directive was issued. When all the necessary changes occur, the CRBE is informed (5) about the necessity to invite a new device and terminate the current session with the other. After, it transforms this request into two distinct actions towards the SM (6). This first (7) refers to the session establishment with the new terminal (B). If it supports multicast, it joins the multicast addresses (8a) contained within the Service Delivery Platform (SDP) information of the invite message. Then, it acknowledges (9a) this process and starts receiving the content (10a). If, on the other hand, the terminal only supports unicast, it sends back a 200 OK message (8b) containing the address requirements (IP and ports) necessary to conclude the overlay trees' reconfiguration (explained in session initiation). Once this message is received by the SM, it is forwarded to the CRBE (9b) and subsequently to the NTL (10b). At this stage, the overlay trees are conveniently updated and the user starts receiving the newly adapted multimedia content (11b). Once this process is finalized, it can be considered that the session is transferred to the new terminal. Nevertheless, it is necessary to end the session previously established with the other device (A). This is done by using a SIP Bye message (12). If the device is multicast enabled, it sends a leave message to all the sub-groups it was associated with (13). Finally, the 200 OK message (14) confirms that everything has successfully terminated. At this stage the SM does not need to notify the CRBE or the NTL because the resources reserved for these sub-groups were initially released (between Steps 4 and 5).

7.5.6 Assumptions and considerations

For the most attentive and critical reader, there might be some questionable aspects within the use-cases. This subsection attempts to expose some of the assumptions and considerations taken when designing the use-cases. In this sense, it is likely that some of these questions can be dissipated.

7.5.6.1 Multimedia applications and advertising

Although the use-cases are mainly related with advertising, the Context-Aware Multimedia Management System can be used without any changes within other multimedia applications. Some of the scenarios presented might not yet be a reality in the realm of advertising, but most of them are already being explored in other fields of multimedia delivery (see section 7.2). In this sense, while some people might consider that handovers or adaptations might not be critical for advertising (because usually the multimedia interactions are very short), it is probable that the advertising industry will change these paradigms. In fact, embedded advertising is already a reality, where products, services or ideas are coupled together with other long-duration multimedia content. In this sense, the proposed solution foresees these and other concerns associated with broader multimedia scopes.

7.5.6.2 Security and privacy

The security and privacy issues have been explained in sections 5.4.3 and 5.5.1. To improve the diagrams readability, all the procedures related with this topic were excluded from the representation. Nevertheless, for all the use-cases it is clear that when setting a campaign, the advertiser sends an authenticated token (confirming his authenticity), which is used by the CRBE. In this sense, the CRBE subscribes to the users context on behalf of the advertiser. Obviously, if the advertiser does not possess the required permissions, the user will not be targeted and the context not requested or subscribed. The same applies for the rules associated with multimedia delivery.

7.5.6.3 Missing interactions

To allow a better understanding of the specifics of each use-case, there are some interactions, which were omitted from the diagrams. The following list contains some examples of the assumptions. Other assumptions or omissions are described in the use-cases.

- The CRBE needs to be previously subscribed in order to receive user-context notifications.
- Some acknowledgment mechanisms are not complete. In some situations it may happen that the 'cascade' of acknowledgements is not done until the last entity.
- The diagrams assume the users have previously authorized the advertisers to target them.
- The diagrams assume the users have previously registered into a know SIP Registrar server.
- The user device must support SIP signalling and multiple media streams.

7.6 Summary and outlook

The convergence of both telecommunications and Internet worlds into a unified IP-based environment, together with the recent advances in broadband access, allowed multimedia communication services to grow like never seen before. Likewise, the adoption of IP-capable multimedia-enabled devices is increasing every day. Altogether, while it may seem great news for the industry, the diversity of solutions resulting from recent technological advances also imposes big challenges.

Just like in any other service, multimedia-based advertising has the purpose to improve customer-perceived QoE, by providing personalized, contextualized and engaging content. However, the heterogeneity of access networks, devices, environments and physical mobility does not facilitate this process. At the same time, the cost of targeting personalization is usually paid by the lack of efficiency.

In this sense, this work proposes a multimedia delivery framework capable of addressing the aforementioned problems. Focusing at the application and session layers, it uses context-aware technologies to achieve a balance between personalization and efficiency. This is achieved on the basis of two fundamental principles. The first is called sub-grouping and consists in allowing a multimedia file to be separated into different media types (audio, video and text) and later stream it in the most appropriate combination media types and versions (encodings). The second is supported by a generic context-aware framework, which enables the selection of the most appropriate encodings based on a series of current context conditions.

Therefore, by using the context management framework defined in Chapter 5 and the reasoning engine methodology introduced in Chapter 6, the SME is capable of triggering multimedia delivery across heterogeneous environments according to a different set of parameters (context) and adapt it in real time, accordingly. Altogether, when combining the work developed in Chapters 5–7 the proposed Converged User-Centric Advertising Framework (CUCAF) is realized. In sequence, the following chapters give an overview of the implementation efforts, as well as expose the validation and evaluation techniques used to test the proposed framework.

Chapter 8
Implementation and testbed description

Following the specification and conceptual design for the advertising platform introduced in Chapter 4 and the detailed architectural descriptions presented in Chapters 5–7, this chapter describes the implementation details for the advertising platform prototype. To facilitate the comprehension of what was implemented, Figure 8.1 depicts a detailed representation of the Converged User-Centric Advertising Framework (CUCAF) functional reference architecture (contains the sub-modules not mentioned in Figure 4.9). The actual implementation of the CUCAF, which is presented in this chapter, is called Converged User-Centric Advertising System (CUCAS).

Section 8.1 overviews what was developed from scratch, what was improved and what was re-used from third parties. Then, section 8.2 provides a specific description of every entity covered by this work while analysing some of its implementation decisions. Lastly, section 8.3 explains how different components communicate with each other, how they were physically implemented and how they were set up for testing and evaluation purposes. Furthermore, it briefly highlights the limitations and dependencies of the overall platform.

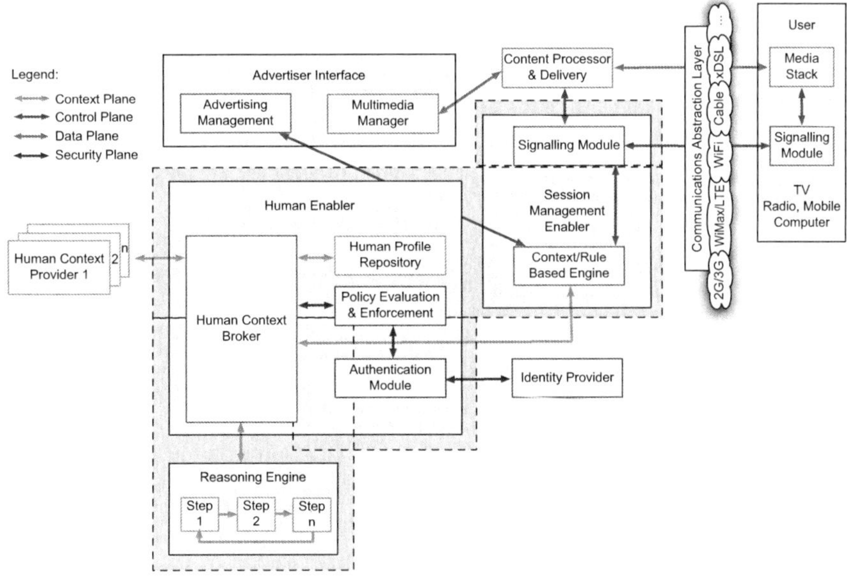

Figure 8.1 Detailed CUCAF functional reference architecture and adjacent entities

8.1 Realization overview

Due to the amount of components developed within the scope of this dissertation, it was not possible to create a single system, developed under a single programming environment. Moreover, as described before, in some situations, the components were completely developed in the scope of this work, while in others, only some extensions were provided. In this sense, it is important to make clear where this book's contributions are most notable and understand why in some cases, other components were re-used or extended.

While the next sections present all the entities in detail, Table 8.1 provides an overview of all the components and respective sub-modules, developed or used under this research work. To get a better perception on how these components relate, please review Chapter 4, in particular, Figure 4.9. In short, four components and one sub-module have been completely developed within this dissertation, while four others were improved by existing software from third parties.

Regarding the programming environments used, these varied according to the component. In this sense, this work was developed with the help of the following technologies: PHP, Java, Flex/Flash, HTML, MySQL and Shell Scripting. Further details are provided in the upcoming sections.

8.2 The components and entities

To allow a better understanding of what was implemented, just like in the previous chapters, the analysis is separated into three main components. Then, some

Table 8.1 List of all the components and sub-modules used or developed under the scope of this dissertation

Component/sub-module	Contribution	Implementation details
Advertising management	Total	Developed from scratch
Multimedia manager	None	Developed under the C-CAST project
Content processor and delivery	None	Developed under the C-CAST project
Human profile repository	None	Developed by UASO
Human context broker	Partial	Extended from UASO Context Broker
Policy enforcement and evaluation	Partial	Extended from FOKUS PEEM component
Authentication module	Partial	Extended from FOKUS GUIDE component
Identity provider	None	Provided by FOKUS GUIDE component
Reasoning enabler	Total	Developed from scratch
Session management enabler	Total	Developed from scratch
User client	Partial	Re-used from C-CAST project
Privacy context provider	Total	Developed from scratch
Social context provider	Total	Developed from scratch

sub-modules are covered in a separate subsection. However, only the ones where we make some sort of contribution (i.e. total or partial) will be considered. The integration of the overall advertising solution is covered in section 8.3.

8.2.1 Generic human enabler

The generic human enabler mainly comprises the Context Broker (CxB) entity developed by University of Applied Sciences Osnabruck (Knappmeyer et al., 2009). This component was developed with Java and PHP, under an HTTP application server, namely JBOSS (http://www.jboss.org/). The module responsible for policy management and security enforcement is the Policy Evaluation, Enforcement and Management (PEEM), developed at Fraunhofer FOKUS (Blum et al., 2009). Also developed in a combination of PHP and Java (Software Development Kit (SDK) 6), it uses an HTTP/SIP (Session Initiation Protocol) application server, namely Sailfin (http://sailfin.java.net/), to expose its capabilities to third parties. Lastly, the authentication module, developed in Java (SDK 6) and running on a Sailfin application server, checks the authenticity of every request that is intercepted by the PEEM component. Altogether, the implementation work on this component consists in providing interfaces between all of the previously mentioned components/modules. Furthermore, it is necessary to provide a way to communicate with external components (e.g. context providers and the session management enabler). Figure 8.2 depicts the interfaces and technologies used to establish communications between different components/modules.

To allow a better understanding of the implementation work, the analysis will be separated by each interface:

- *I-A*: used to perform historical queries. This is part of the original CxB, and therefore, this dissertation has not contributed to this interface.
- *I-B*: due to the design constrains on the CxB, which support REST only, together with the Simple Object Access Protocol (SOAP) constraints on the PEEM, it was necessary to create a module inside PEEM that was capable of

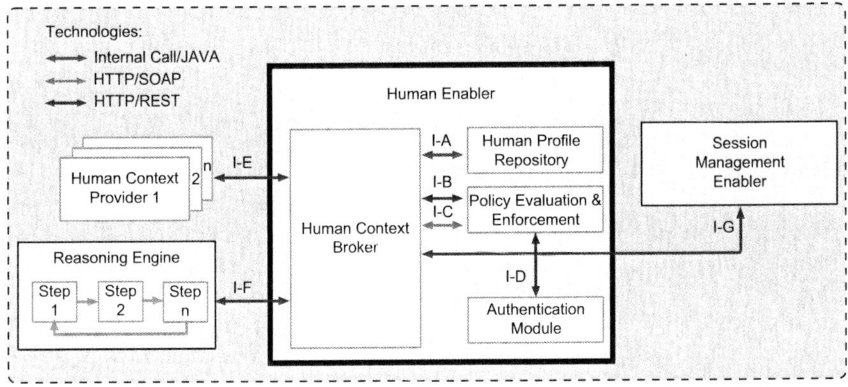

Figure 8.2 Interfaces and technologies used within the generic human enabler

requesting additional context information to perform evaluation and enforcement. In this sense, this work contributed with a module capable of invoking REST services.

- *I-C*: similarly to I-B, but on the inverse direction, whenever a new request, subscription or notification occurs, the CxB needs to query the PEEM to validate these actions. As the PEEM only supports SOAP requests, an extra module was inserted within the CxB and the functionalities extended to support security and privacy. In other words, requests and subscriptions towards the CxB can, optionally, take into account user privacy settings.
- *I-D*: as the PEEM was originally designed to support authorization and not authentication, it was necessary to extend its functionalities. Therefore, this simple interface was created to make sure the requests are authentic. The module just checks if the hashes are ok and validates (i.e. authenticates) the request.
- *I-E*: nothing was implemented here. However, both the social and privacy context providers had to follow the CxB specifications to support ContextML over HTTP/REST.
- *I-F*: similar to the above (i.e. I-E). All information that was generated from the reasoning enabler was transformed into the ContextML format, allowing it to be managed by the Human Context Broker.
- *I-G*: this interface exists to request, subscribe and notify the HCB and SME, respectively. Again, the protocol used is REST over HTTP.

8.2.1.1 Implementation decisions

Based on the requirements initially defined in section 3.3.1 and taking into account the state-of-the-art of user-data management solutions, our strategy opts for extending the model that was found most appropriate. The main reason for a context-aware brokerage system and the ContextML language relates to its flexibility and extensibility. This allows the system to adapt to any type of user-related data (i.e. context). Furthermore, the brokerage system has proved to be scalable by using a distributed/federated architecture (Gomes et al., 2010).

Nevertheless, security and privacy was not covered under these systems. Due to the scenarios that were anticipated for the management of such advertising platform (e.g. managed by a telecom operator), it was necessary that policies assuring privacy and security were compliant with industry standards. Furthermore, it needed to be flexible enough to allow other scenarios. For these reasons, the decision went to the FOKUS PEEM implementation, which follows Open Mobile Alliance (OMA) specifications. Moreover, it was a component already available at existing playground initiatives. However, it would probably be more scalable and efficient to use Drools (http://www.jboss.org/drools) (or other similar products), a business logic integration platform, that could be used for the same purposes.

In what concerns the authentication strategy, the proposed concept assumes a circle of trust between the authentication module and the identity provider. It seemed a very valid approach because nowadays, mostly any service allows some sort of Single Sign-On (SSO) (e.g. OpenID http://openid.net), authentication mechanism. In fact, this is already a common practice among different, but federated domains (e.g. Internet, telecommunications or social networks).

8.2.2 New knowledge prediction engine

Given the fact that the reasoning enabler is not exactly an architectural component, but more an implementation of a methodology, its implementation is not so straightforward. As opposed to the previous user-related data, which was managed by the Human Context Broker, in the carried experiments, for simplicity reasons, the information was extracted directly from the Social Context Provider's database. For this purpose, a MySQL server (http://www.mysql.com) was used. All the pre-processing (i.e. cleaning data and preparing it for mining) steps were implemented, either by SQL queries or directly with Java (SDK 6) code.

In what concerns data mining techniques, the option went for clustering. Nevertheless, as explained thoroughly in Chapter 6, there are different types of clustering being applied. For one-dimensional features (i.e. simple context types), the Density-Based Spatial Clustering of Applications with Noise (DBSCAN) algorithm (Ester et al., 1996) was used, supporting both Euclidean and Manhathan (slightly modified) distances. These changes have been done in JAVA (SDK 6) using the Weka (http://www.cs.waikato.ac.nz/ml/weka) library. As for the multi-dimensional clustering, previously called, Consensus Clustering, the Self-Organizing Map (SOM) algorithm (Kohonen, 1997), was applied and implemented within the GenePattern framework (Reich et al., 2006). Despite these steps are independent from each other, a shell script together with all the Java (SDK 6) libraries was developed to operationalize the entire process. Once the entire clustering operations were finished, the different sets of algorithms (developed in Java) would run. Again, the logic was built-in Java (SDK 6). Then, all results were transformed into a ContextML format so that it could be published into the HCB. To allow a better understanding of which technologies were used, under each step, Figure 8.3 depicts the implemented solution.

In Figure 8.3, the dotted lines separating each individual step are then connected by a shell script, which is responsible for controlling the outcome of each step and reuse the obtained results to the following. The Operating System (OS) used was Linux, nevertheless it could run on any other that supports Java. In this case, only the shell script would have to be replaced for an equivalent, in the chosen OS.

Despite using Weka as a data mining solution, there were other valid options, such as the R Project (http://www.r-project.org) or Rapid Miner (http://www.rapid-i.com) (both open-source). Due to the narrow scope of the implementation, the decision went for Weka for two main reasons. The first was that it had previous experience with this tool. The second was that it contained all the needed algorithms plus the necessary sources to modify them when it was required (e.g. Manhattan distance). Nevertheless, in the future, to allow a vast range of data mining options, it would probably be best to use the R Project, which has a very solid community, complete documentation and its architecture lays on a very flexible modular approach. Therefore, it is easily extendible and consequently, whenever it is necessary, a new algorithm or plug-in can be easily added.

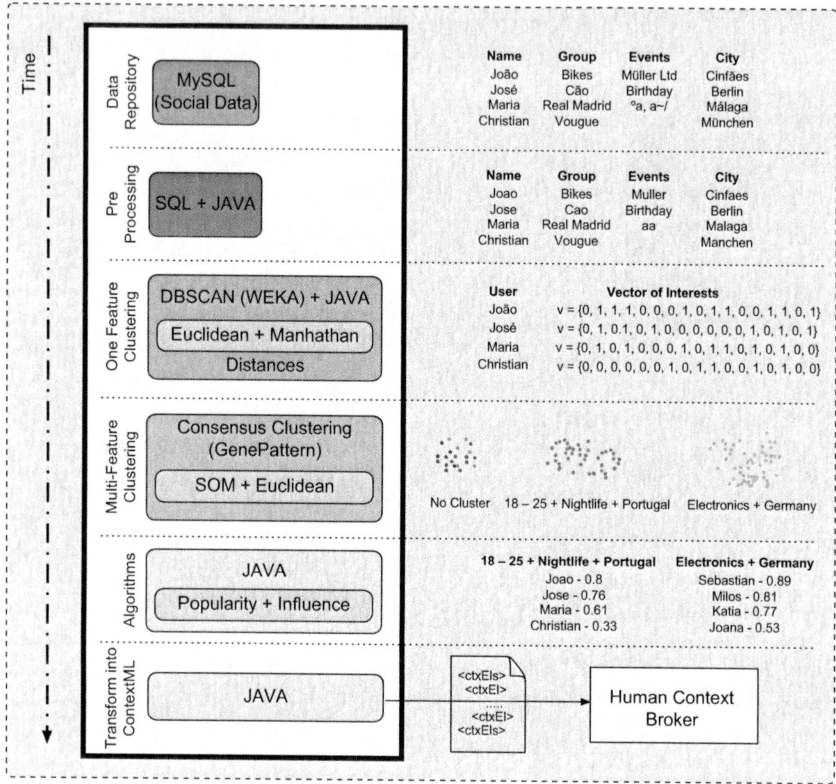

Figure 8.3 Technical details of the reasoning enabler implemented

8.2.3 Session management enabler

The session management enabler was entirely developed in Java (SDK 6), but in two complementary modules. Both were compiled as Web application ARchive (WAR) files, and deployed in a Sailfin Application Server (supports both SIP and HTTP). Figure 8.4 gives an overview of all the components, modules and sub-modules associated with the SME.

Although the interfaces between the different components and modules have been described in Chapter 7, in some cases, the technologies used to exchange information were not mentioned. In this sense, Table 8.2 provides the details to the aforementioned.

In order to enable third-party applications to make use of this enabler, the supported capabilities are exposed through open programmable APIs, specified with the Web Service Description Language (WSDL). The realization of the related web services is based on the Apache EXtensible Interaction System (Axis) libraries (http://axis.apache.org), which is an open-source implementation for the web service communication protocol SOAP. SOAP is a lightweight protocol for exchanging structured information in a decentralized, distributed environment. Axis is a

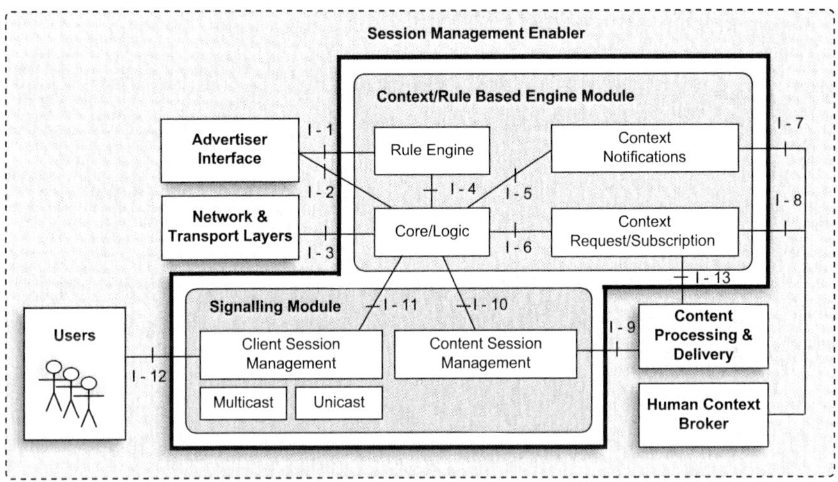

Figure 8.4 Implemented SME modules and respective interfaces towards other components

SOAP framework for constructing SOAP communication entities such as clients, servers, gateways, etc. In addition, the platform Java EE (Java Platform, Enterprise Edition version) is used for the realization of web services and SIP Servlets. The related components are merely deployed in particular containers that implement the Java EE specification.

More specifically, the SME comprises two entities, the Context/Rule-Based Engine (CRBE) and the signalling modules. The next subsections will allow a better understanding of the implementation details and decisions, for each of them. Basically the first deals with the logic and external interfaces, while the second takes care of the multimedia sessions.

Table 8.2 Description of the technology used for each interface within the SME

Interface	Component/sub-module	Technology
I-1	AI-RE	HTTP REST Interface
I-2	AI-Core	Web Service (SOAP)
I-3	NTL-Core	Web Service (SOAP)
I-4	RE-Core	PEEM-1 and REST Interfaces
I-5	Core-CN	Internal code (Java)
I-6	Core-CR	Internal code (Java)
I-7	CN-HCB	HTTP REST Interface
I-8	CR-HCB	HTTP REST Interface
I-9	CSMa-CtPD	SIP signalling
I-10	Core-CSMa	Web Service (SOAP)
I-11	Core-CSMb	Web Service (SOAP)
I-12	CSMb-Users	SIP signalling
I-13	CR-CtPD	HTTP REST Interface

8.2.3.1 Context/Rule-Based Engine Module

The best way to understand what was implemented is by overviewing the different sub-modules in action. In this sense, Figure 8.5 depicts the activity flow within the CRBE for a session initiation process. In fact, all other working modes (i.e. session modification, termination, etc.) will fit in a sub-set of these states.

The module is always activated through some sort of web service (I-2) invocation (explained earlier), and the supported methods include: creation of advertising groups (i.e. targeted ads), add users, remove users and terminate group. The creation of an advertising group (or campaign) can be configured as static (i.e. known list of users) or dynamic (i.e. users selected according to the context types defined in the advertising interface).

Once the advertisement arrives the module, the XML message is parsed, stored in a local MySQL database and the triggering rules are created/updated in the PEEM component (I-4). As this component can be used for policy evaluation and enforcement, in this step, it will be acting as a rule engine. Afterwards, the logic is responsible for subscribing and requesting (I-5, I-6, I-7, I-8) all the necessary context types (i.e. towards the CtPD and the HCB). This process can take more than a single loop when the users to be targeted are not initially known.

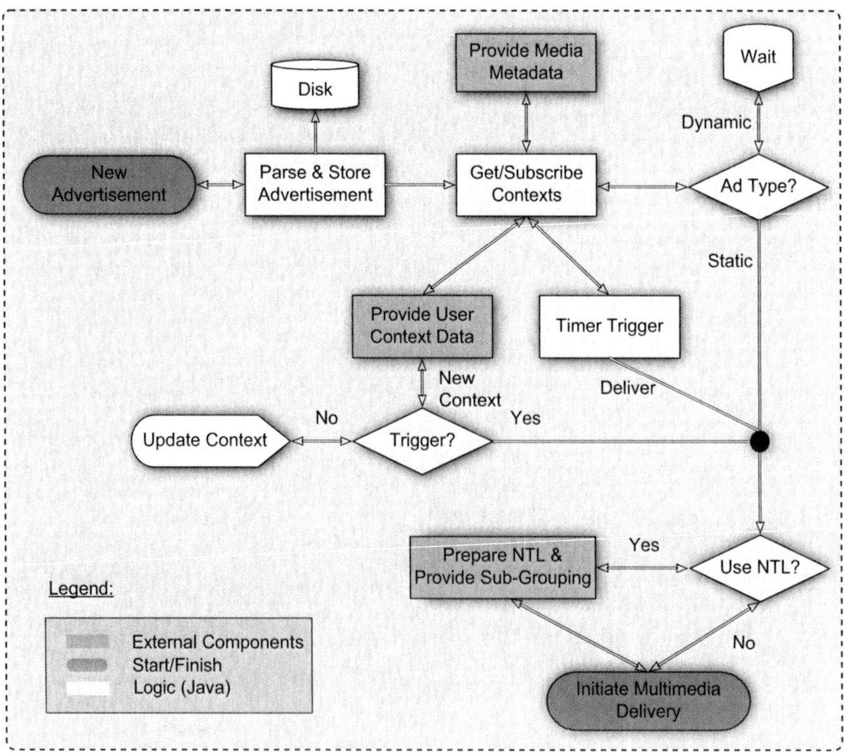

Figure 8.5 Context/rule-based engine activity flow diagram

```
<contentDescription>
  <mediaTypes>
    <media type=" audio ">
      <codecs>
        <codec type="MP3">
          <availableSamplerates>
            <samplerate value="44100"/>
            <samplerate value="48000"/>
          </availableSamplerates>
        </codec>
        <codec type="A52">
          <availableSamplerates>
            <samplerate value="32000"/>
            <samplerate value="48000"/>
          </availableSamplerates>
        </codec>
      </codecs>
    </media>
    <media type=" video ">
      <codecs>
        <codec type="MP2">
          <availableResolutions>
            <resolution value="320x480">
          <availableFramerates>
              <framerate value="50">
            <availableBitrates>
                <bitrate value="3000000"/>
            </availableBitrates>
              </framerate>
          </availableFramerates>
            </resolution>
            <resolution value="640x352">
        ......
        .....
      </codecs>
    </media>
  </mediaTypes>
  <status code="0" description=" "/>
</contentDescription>
```

Listing 8.1 Content description exchanged in the interface between the SME and the CtPD

Specifically for the content context (i.e. description), there is an interface (I-13), which describes what kind of media types are available for each content (in this case, advertisement). Based on this meta-description, the SME is capable of performing sub-grouping and consequently adapt the multimedia streams, according to the user's preferences or device's capabilities. Listing 8.1 provides an example of the content metadata exchanged between the CtPD and the SME.

Then, if the trigger type is time, an internal sub-module is activated (i.e. new thread is created). The latter is responsible for triggering a notification when the timer reaches zero. Alternatively, if the trigger type is another context source, the delivery only occurs, when a notification matching the initial trigger values is met. On the other hand, if the advertisement type is set to static, the delivery is promptly activated (after getting the initial user and content context information).

When available, the interfaces towards the network and transport layers are invoked (I-3). This allows the overall network to be re-arranged, ensuring end-to-end quality of service. In this case, those layers and respective components are the ones responsible for the sub-grouping operations. Otherwise, the CRBE is in charge of matching the available codecs (i.e. from the content) with the ones supported by the user's devices and preferences. Finally, when all the user's data, necessary for delivery, is known, the signalling module is invoked (I-10 and I-11).

8.2.3.2 Signalling module

The signalling module applicability begins where the CRBE's finishes. At this point, all the users and the respective media format to be streamed are known. Basically, the functionality of this module is to act as a B2BUA. Figure 8.6 highlights this process. On the one hand, for each user, a SIP INVITE message is generated (I-11). This message includes the media type to be streamed, the IPs and ports from the streamer (CtPD), as well as some parameters, which will be used for synchronizing different media types. On the other hand, a similar process is initiated towards the CtPD (I-10). However, in this case, a single session is created. The

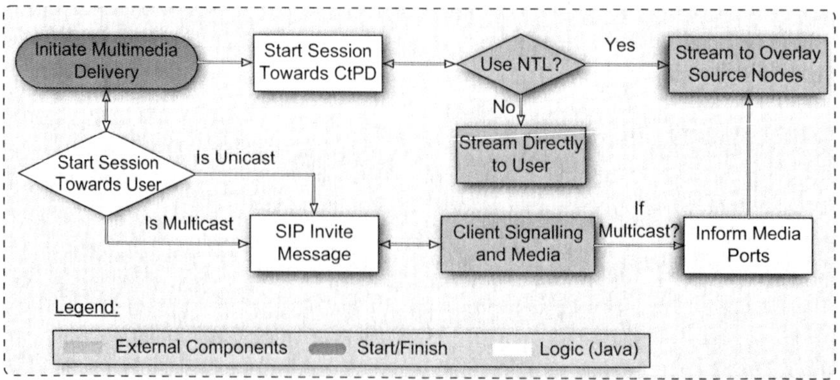

Figure 8.6 Signalling module activity flow diagram

details of the remaining scenarios will not be described as these have been thoroughly described in section 7.5. Instead, the following paragraphs highlight some of the implementation decisions taken, to enable the execution of the aforementioned scenarios.

One of the biggest challenges was separating multimedia content into different media types (i.e. audio, video and subtitles). Although most of the work is done at the CtPD, where the media is separated, it is also important to signal that information to the client (further details are explained ahead in a separate – client – subsection). Moreover, as mentioned previously, the CtPD was a component developed under the C-CAST project, which was re-used to test streaming capabilities for the advertising solution.

In order to support new users to join existing media sessions, due to the limitations on the CtPD, which could not start streaming from a specific point in time, some tweaks were applied to overcome this situation. In this sense, all media types (for the same content) would have to be initialized from the beginning. To minimize the overhead of this approach, the SME would instruct the CtPD to start streaming to a '/dev/null' point (in the local – CtPD – machine), where no network resources would be used. Alternatively, the SME could indicate that these media types (the ones not being streamed to the users) should be streamed to an overlay source node (or any other node), where the packets would eventually be discarded as no user belonged to that overlay tree.

Please note that this would reduce the computing overhead on the CtPD machine, and the network overhead would not be so significant, because this would occur at a local network level. Just like on the previous sub-module (CRBE), the web services exposed here have exactly the same purpose as before. The main difference is that the previous relates to the logic, while these to the execution of the actions towards the users and content provider.

8.2.4 Other components

To achieve the initially proposed advertising architecture, some additional components were developed, or re-used, to enable the full capabilities of the solution being proposed. Therefore, this section focuses on the client, advertising management, as well as privacy and social context providers.

8.2.4.1 User client

Although one of the main objectives in the proposed framework was reducing the number of changes in existing hardware and software throughout the entire value of chain, the reality showed that not many devices (particularly mobile) are prepared to support SIP (mandatory in such solution), which is the most used protocol for signalling, over IP networks. Furthermore, they should support RTP to stream multimedia. By overviewing the state-of-the art of existing mobile terminals, the choice went to a model based on the Android (http://www.android.com/) OS. At this point in time (software version 2.0), there was no SIP stack available; therefore, the MJSIP stack (http://www.crealabnet.it/download/MjSIP_ENG.pdf) was

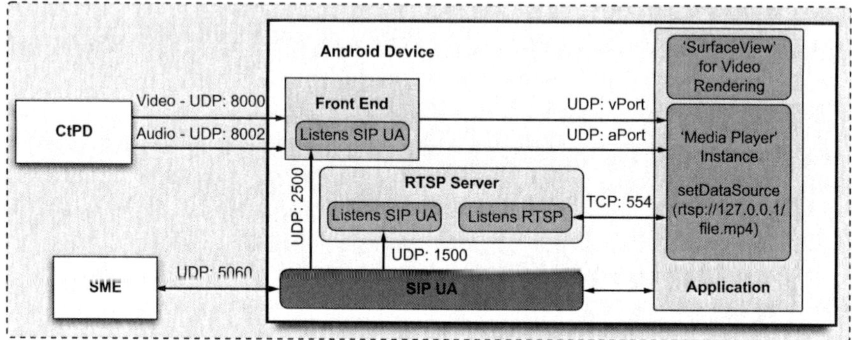

Figure 8.7 Client architectural design

ported into the platform. Moreover, the device media player only supported Real-Time Streaming Protocol (RTSP), while the architecture used SIP together with RTP. Therefore, it was necessary to tweak the system. For this purpose, an RTSP server was built-in on the device, and a specific sub-module was developed to make the necessary adaptations from SIP plus RTP, to RTSP. Figure 8.7 provides the schematic representation of the engineered solution.

In short, multimedia is signalled from the SME to the SIP client, installed on the Android device. After receiving it, it parses the media descriptions inside the SDP and triggers the application to make a request on the media player (RTSP client). Consequently the player opens a fake URL, such as, rtsp://127.0.0.1/file.mp4. Then, the RTSP server receives the client request and on its answer, it provides the UDP ports that the player should listen to (alternatively, the player could choose the ports and the server stream to them). As the front end listens to the UDP ports that were agreed with the SIP signalling. Internally, it forwards the packets to the ports defined by the player. In this implementation, these ports were static. However, in the C-CAST implementation, an extra component (MTO reliability) within the client terminal, would, among other things (e.g. handovers), allow Forward Error Correction to be used. It is also important to notice that this mechanism works for both unicast and multicast transmissions.

8.2.4.2 Privacy context provider

Despite the amount of interactions involved around this component, its conception is rather simple. Its interface is built in a Flex/Flash web application, in combination with an HTTP Servlet, running on a Sailfin application server. Figure 8.8 provides some details over the implementation and respective interfaces towards other components.

The first step involves the user authentication (depicted in Figure 5.5). Then, the user can access (I-1) its groups (user's lists) and manage them accordingly. Similarly, he/she can retrieve all the possible context types and define policies accordingly. The context types are fetched directly from the HCB (I-2), by invoking its REST web services (from the Flex application). Whenever there is a change

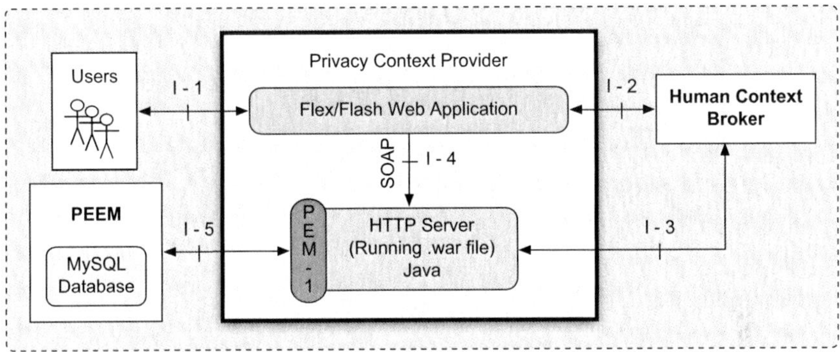

Figure 8.8 Privacy context provider architectural representation

(i.e. lists or policies), this is sent (I-4) to the application running on the Sailfin container. The main reason why the communication does not occur directly with the PEEM component relates to the fact that jar files cannot be appended (embedded) to Flex web applications. In this sense, this component acts as a mediator between both parties. The communication between the PEEM and the HTTP Servlet (I-5) is based on the PEM-1 interface (specified by OMA). For this purpose, a library already existing at Fraunhofer FOKUS SOA Playground (http://www.opensoaplayground.org) was used. Additionally, by using interface I-3, the policies can be replicated to the HCB under the ContextML format.

Despite the fact that in Chapter 5 a very flexible model for policy control was introduced, in practice, not all features were implemented. Nevertheless, it provides most of the privacy control settings available in the most common web-based services. In addition, it extends the concept by providing the means to set up real-time context-aware policies. Figure 8.9 provides a screenshot for this situation.

However, the most usual way of defining policies, nowadays, is through connections (e.g friends) and lists. In this sense, this component allows users to create connections with both other users and lists of users. This simplifies the process of policy management. In fact, due to the nature of the FOKUS PEEM implementation (the one being used in the implementation), group/list management already belongs to its core functionalities. Figures 8.10 and 8.11 provide a sample of what is possible to achieve with this implementation.

8.2.4.3 Social context provider

The social context provider is mainly a set of applications that gather information from online social networks and store it under a unified structure. Within this work, two versions for this component were developed. The first focused on the data integration coming from two distinct social networks (i.e. Facebook (http://www.facebook.com) and Orkut (http://www.orkut.com)), while the second on a single network (Facebook – Social Enabler (http://apps.facebook.com/socialenabler)).

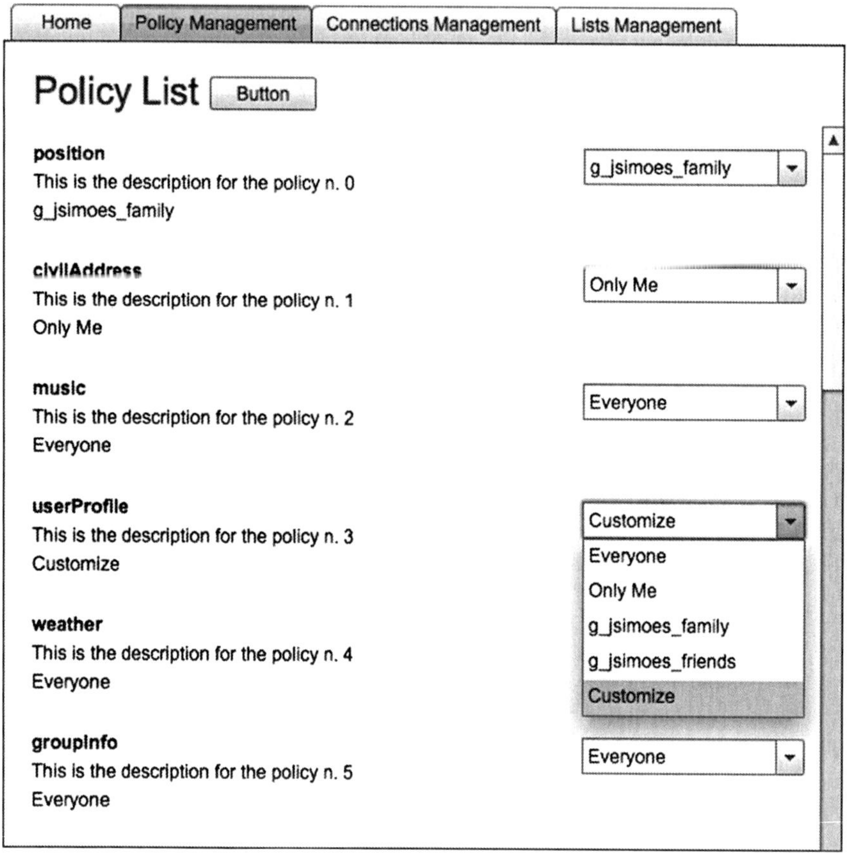

Figure 8.9 Screenshots of the privacy context provider user interface

Nevertheless, both of them used the same architecture. Both the logic and the application that collects information about the users were developed in PHP. The main reason for this relates to the fact that these social networks provided their APIs over this technology. To store data a MySQL database was used. The initial algorithms were very simple formulas executed in PHP. Nevertheless, afterwards, due to the amount of information available, data mining technologies were used to provide more accurate information (details can be found in Chapter 6). Further details of this work can be found in Simoes et al. (2010).

As a result of the first implementation efforts, a social data visualization tool was developed. It was implemented by using the RAVIS (http://code.google.com/p/birdeye/wiki/RaVis) framework, a complex data visualization library for Flex. All the communication between the Flex social graph component and the logic back-end was done by HTTP web service calls and XML descriptions. In addition, by using the ZEND (http://www.zend.com) framework, the PHP functions were exposed to any third-party service (i.e. the HCB). In total, more than 25 different

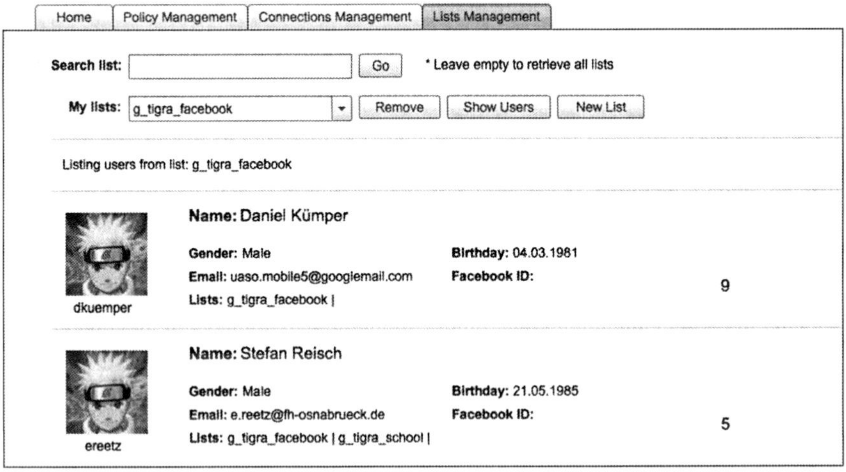

Figure 8.10 Screenshots of the privacy context provider: list management

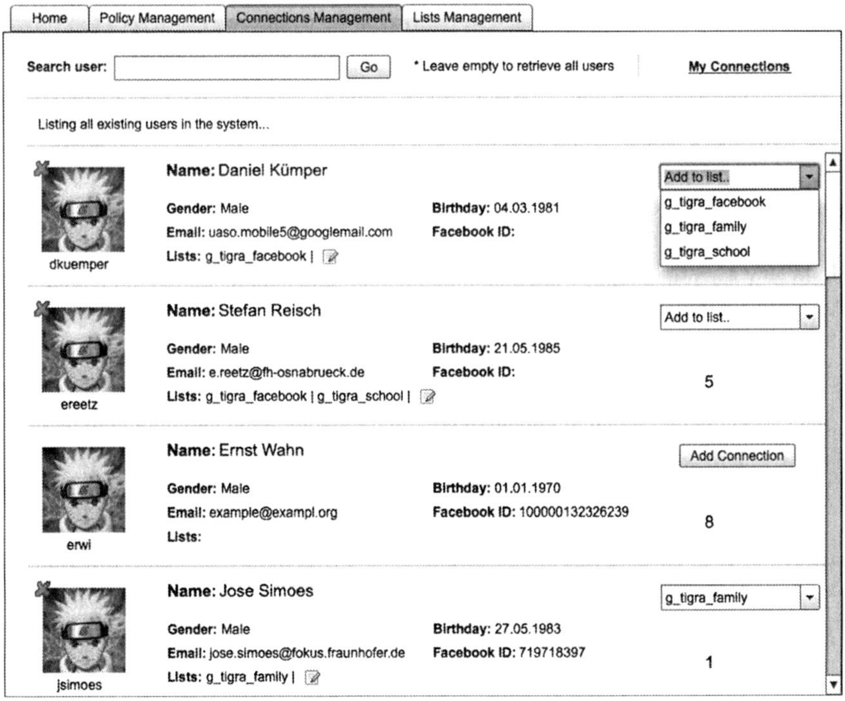

Figure 8.11 Screenshot of the privacy context provider: policy management

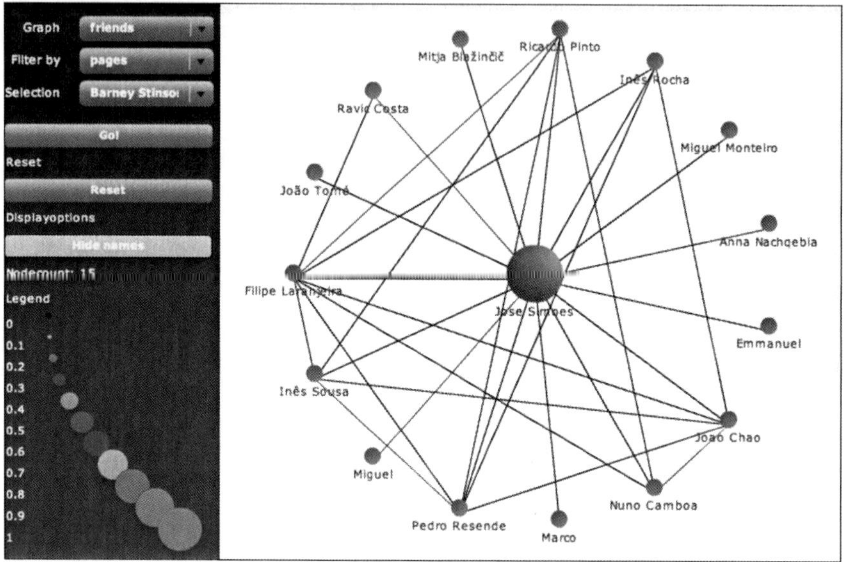

Figure 8.12 Screenshot from the Facebook social application (similarity by interests)

methods were made available. These ranged from simple data queries to the results of complex algorithmic operations (e.g. popularity). To allow a better perception of what was viewed (experienced) by the end-users, Figure 8.12 shows a screenshot from the Facebook social application.

8.2.4.4 Advertising management

The advertising management is mainly built as a Flex/Flash web application. In short, the advertisement setup is done using a graphical interface, which invokes a web service, available on the SME (Core/Logic). The remaining process is described in section 8.2.3. Figure 8.13 depicts some of the possibilities given to advertisers to set up a new advertisement.

Due to the fact that the main purpose of this implementation was testing a concept and not developing a product, not all contexts were made available at this component. In fact, only a subset of them was implemented, more concretely: user demographics, time, location (position), weather and presence status. For testing purposes, all of these context providers were re-used from the work done in the C-CAST European project. Nevertheless, despite limited, the prototype offers much more options than other works. Below is a description of the possibilities for each context type:

- *Demographics*: allows specifying age, gender and some basic interests. Nowadays, most solutions do not allow better targeting than this.

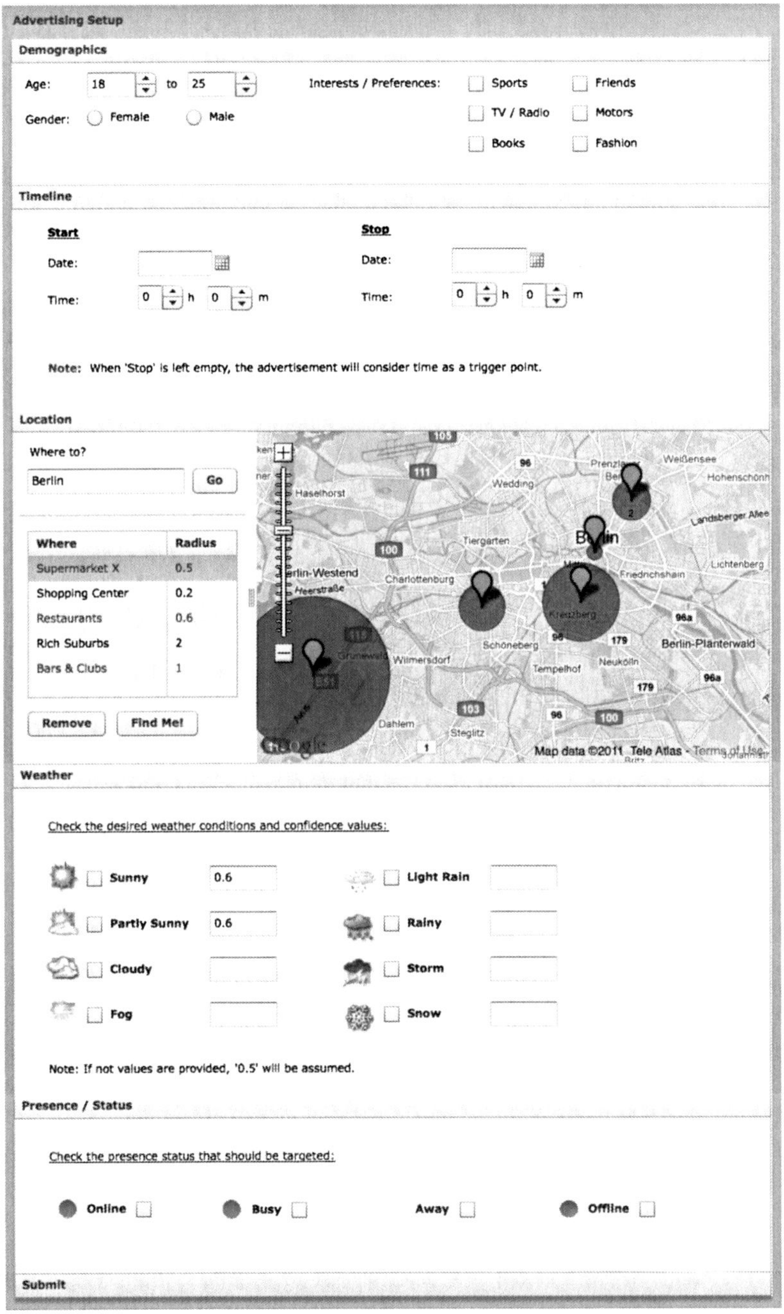

Figure 8.13 Screenshots of the advertising management graphical interface

- *Timeline*: there are two types of options: campaign and action. While the first establishes the time period under which the advertisement should be active, the second indicates that the ad too should be experienced at a specific point in time.
- *Location*: probably the most valuable context type for advertisers. It allows specifying directly in a map which points should be targeted for a campaign. Furthermore, it enables setting up the radius of the targeting for each specific place. To simplify the task, advertisers can search for specific locations. This is ideal for a multi-store promotion.
- *Weather*: one of the most interesting context types. In here, it is possible to specify, which weather conditions should trigger the advertisement. Moreover, it is possible to indicate the degree of certainty for each condition (e.g. 0.8 means there is 80% of probability for a specific condition to occur).
- *Presence*: most users will likely disregard an advertisement if their status is set as 'busy'. Similarly, if they are 'away', they will probably miss it. In this sense, this option allows advertisers to work around these situations.
- *Delivery*: represents the interface where it is possible to choose the way content will be delivered. Advertisers can specify a list of users, or allow automatic context-aware delivery. Furthermore, they can choose between a specific multimedia content and SMS delivery.

Obviously, all of the aforementioned features only represent a fraction of what is possible. For example delivery methods could be more personalized (e.g. if mobile supports MMS the advertisement will arrive in the form of a picture, otherwise, in text). A combination of other contexts is also achievable, although some modifications would be required in both the interface and the logic (set of rules) on the SME. Nevertheless, these topics are further explored in Chapter 10.

8.2.4.5 Other minor implementations

Finally, there were two other implementations within this work. The first one relates to the authentication module, which is represented by a function that checks the integrity of the token. The code was developed in a shell script and used the OpenSSL library (http://www.openssl.org/). Alternatively, this method could be implemented directly on the PEEM as an internal function (set of operations).

The second implementation effort was done on the PEEM entity from FOKUS. As the component was already existent and possessed all the necessary capabilities for this project, it was decided to extend its functionalities to accommodate with the interconnection requirements of other components (namely, the HCB, which does not support SOAP). Therefore, a new module was created, allowing the enforcement of services based on the REST protocol (before, only SOAP was supported). In addition, a couple of functions/operators were added to improve the granularity of the rules (e.g. boolean: is in the vicinity?) to be used by the advertising management interface.

8.3 Enhanced advertising testbed description

Until here, all the components, modules and sub-modules were introduced as isolated entities. This section shows how they are used together to achieve the aforementioned, enhanced advertising solution. Furthermore, the implementation details, such as network configuration, computational characteristics (e.g. CPU, memory), libraries or programming language used, will be detailed for each architectural constituent.

The testbed was set up at Fraunhofer FOKUS premises. In total, it used 11 machines, of which, 6 were running as virtual machines. Still part of the testbed, but outside the same premises, was the human context broker, as well as the context providers, which were hosted in UASO (http://www.hs-osnabrueck.de), Germany. The devices were connected to the local network through a wireless access-point. Regarding the devices, all tests were done using Android smart phones. Within the tests, the following models were tested: HTC Hero, HTC Desire, Google Dev Phone 1, Google Nexus and HTC Magic.

To allow a better understanding of how the testbed was composed, Figure 8.14 provides an overview of the components involved, while the following subsection describes each of the entities in more detail.

8.3.1 Component configuration

This section details how each of the entities introduced earlier was implemented in the testbed. Whenever the host running represents a Virtual Machine (VM), its properties will also be described. Moreover, since each of the VMs has different hardware and software requirements and no guest VM needs more than 4GB of Random Access Memory (RAM), it was decided to stick with 32-bit guest operating systems, when possible.

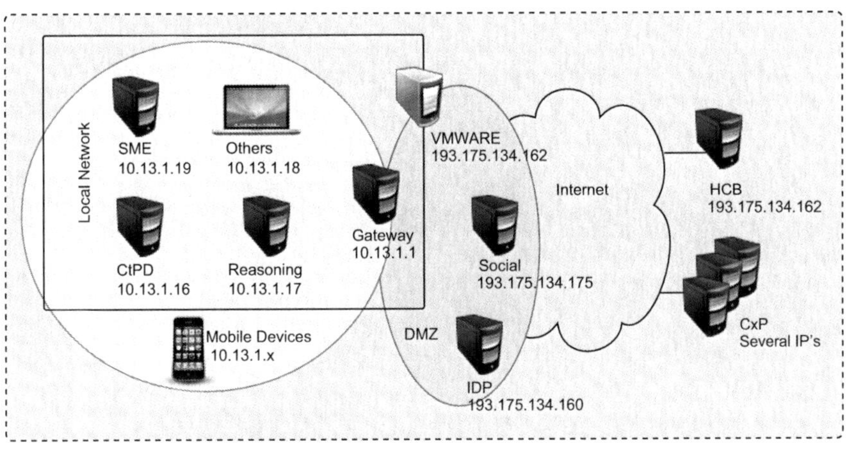

Figure 8.14 Advertising solution testbed description

8.3.2 Session Management Enabler (SME)

IP Address: 10.13.1.19
Operating System: Ubuntu Server 9.04 32bit
Running: Context/Rule-Based Engine, Signalling modules, Open IMS Core
System Configuration: VCores: 2 RAM: 1GB HD: 20GB

8.3.3 Content Processor and Delivery (CtPD)

IP Address: 10.13.1.16
Operating System: Ubuntu Server 9.04 32bit
Running: runs natively on the OS
System Configuration: VCores: 2 RAM: 3GB HD: 20GB

8.3.4 Reasoning Enabler (RE)

IP Address: 10.13.1.17
Operating System: Ubuntu Server 9.04 32bit
Running: WeKa, GenePattern, Java SDK6
System Configurarion: VCores: 2 RAM: 4GB HD: 60GB

8.3.5 Others

IP Address: 10.13.1.18
Operating System: Mac OS X - Snow Leopard
Running: PEEM, Authorization module, Privacy Context Provider, Advertiser Interface
System Configurarion: 2.53 GHz Intel Core 2 Duo RAM: 4GB HD: 160GB

8.3.6 Human Context Broker (HCB)

IP Address: https://mobcom.ecs.fh-osnabrueck.de/CPS/contextbroker/Context Broker/Operating System: Linux
Running: Java SDK6, JBOSS Application Server Human Profile Repository
System Configurarion: xxx RAM: xGB HD: xGB

8.3.7 Social Context Provider (SCP)

IP Address: 193.175.134.88
Operating System: CentOS 5.1 64bit
Running: PHP libraries, Apache 2, MySQL 5.0
System Configurarion: VCores: 2 RAM: 4GB HD: 20GB

8.3.8 Identity Provider (IdP)

IP Address: 193.175.134.160
Operating System: CentOS 5.1 64bit
Running:
System Configurarion: VCores: 2 RAM: 4GB HD: 20GB

8.4 Summary

Providing evidence that this architectural model works, this section showed how different components were implemented, which technologies they used, and how the interfaces towards sub-modules or other components were realized. Furthermore, it highlighted some of the implementation details and decisions. Based on this, it identifies what was implemented from scratch, what was modified, improved or just re-used from other entities.

However, as denoted throughout this chapter, some of the functionalities previously described were not implemented. The main reason for this relates to the implementation efforts required, which are not justifiable for a prototype (i.e. for demonstration purposes would not bring innovation). Still, this does not invalidate the theoretical descriptions in the previous chapters. Likewise, despite the fact that in section 4.3.3, alternative realizations of the CUCAF have been presented (e.g. enablers acting as standalone entities), the implementation only accounts the full-featured advertising solution, where all the previously described components are combined together to achieve the desired CUCAS.

Chapter 9
Validation and evaluation

After overviewing the problems, the requirements, the design, the specifications and respective prototype implementation, it is important to evaluate the impact of this work in the context of advertising and related industries. In this sense, this chapter presents the evaluation and respective validation mechanisms, as well as a discussion regarding this work's contribution in comparison to other approaches.

9.1 Introduction

Due to the fact that the main purpose of this research was not building an advertising solution, but instead, to introduce a framework that could improve the way advertising is perceived, the prototype implementation focused more on the functionalities and less on efficiency or scalability issues. In other words, a priority was given to functional requirements over the non-functional requirements. The main reason for this decision relates to the fact that those are the most prominent to improve the overall advertising experience.

Nevertheless, despite the work done relies on a prototype implementation, there are minimum requirements (even non-functional) that need to be assured. Therefore, throughout the following sections, all types of requirements will be validated and evaluated when possible. Although there are no benchmarks available, and consequently a comparison is hard to achieve, this work will publish the obtained results in the hope they can be used as a benchmarking metric for upcoming initiatives.

The remaining chapter is organized as follows. The evaluation of the converged user-centric advertising system is covered in section 9.2, where a detailed set of metrics and test scenarios is described. Section 9.3, through the validation of the functional requirements, shows which of the solutions/works address the problems previously identified. This analysis is done at a domain by domain level. On the other hand, section 9.4 deals with the technical aspects of the implementation, namely with the parameters related to the non-functional requirements. This is achieved by the analysis of the results obtained in section 9.2. Then, section 9.5 highlights how the aforementioned results contribute to the improvement of the user requirements. Finally, section 9.6 summarizes our work.

9.2 Evaluation of the CUCAF

Before getting deep into the validation of the requirements specified in Chapter 3, it is necessary to evaluate the overall Converged User-Centric Advertising Framework (CUCAF). Just like the remaining work, this analysis will be performed by domain/enabler. In each of them, a list of the most relevant tests and hypotheses will be described together with the adopted methodology. Then, the results are presented and a brief interpretation is provided.

9.2.1 Evaluation of the human enabler

Despite the fact that this work re-used an implementation of the UASO (University of Applied Sciences of Osnabruck) context broker (central component of the human enabler (HE)), it is important to assess some general metric regarding its functioning. However, it will be more interesting to evaluate how some of these values vary, with the functional (privacy and security) improvements. In this sense, the following list of questions needs to be answered:

- What is the end-to-end delay for context requests and notifications?
- How do these vary when authentication and authorization are introduced?
- What is the implication of complex user self-defined policies?
- Does the number of existing rules affect the system?

To better understand the specificity of each of the aforementioned experiments, the following subsections introduce the methodologies adopted, as well as the obtained results.

9.2.1.1 What is the end-to-end delay for context requests and notifications?

Although there are many functionalities associated with the HE, the most critical ones are context requests and subscription/notification. The main reason for this assumption relates to the fact that context providers are responsible for updating context, and therefore, this process runs under a more or less controlled environment. On the other hand, the system is designed to allow several CxCs to access simultaneously to different types of information (context).

Therefore, to allow a complete understanding of the HE behaviour, several tests would have to be made, as there are innumerous factors that can affect its performance, namely:

- *Type of context scope*: simple vs. complex (depends on a simple context, i.e. the weather depends on the location).
- *Type of request*: simple (1 scope, 1 user), multiple (many scopes or many users), complex (a query is performed involving different scopes or users), publish/subscribe.
- Number of scopes, context providers, validity of context and frequency of update.
- Number of users on the system, among others.

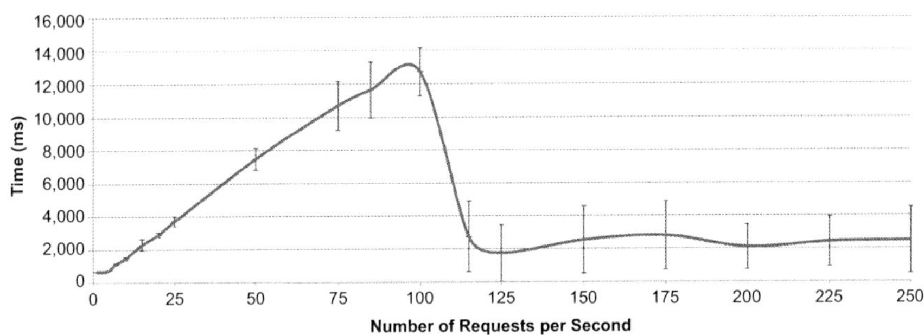

Figure 9.1 Load test to the human context broker for a simple context request (one user)

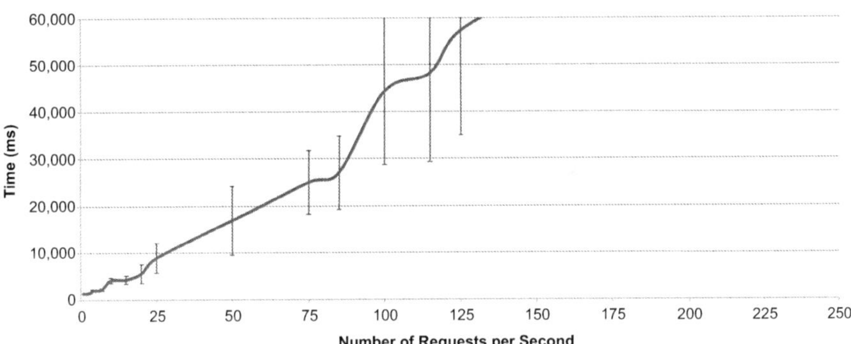

Figure 9.2 Load test to the human context broker for a multiple context request (three users)

To enable a realistic scenario, but without going too deep into the myriad of possibilities, the evaluation of the system will be based on assessing how it reacts when the number of requests per second (req/s) varies. In this test, two different types of requests (one simple and one multiple) are performed, whereas the context scope is always simple. Furthermore, for all tests, the same users were considered. In what concerns context scope, 'userProfile' was chosen, as it is the one that is more often requested by most applications. To perform the aforementioned tests, the Apache Benchmarking[1] tool was used. To make sure that the system was not tested on single bursts of traffic, the number of requests was always 10 times bigger than the concurrency level (e.g. for 5 req/s, 50 requests are sent, therefore the test is performed 10 times). This is also important because it provides average results for each experiment. To better understand the results, Figures 9.1–9.3 show the mean

[1] http://httpd.apache.org/docs/2.0/programs/ab.html

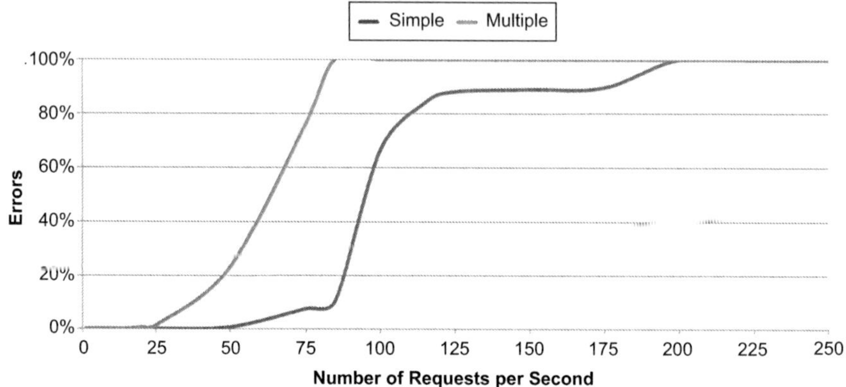

Figure 9.3 Percentage of errors associated with the human context broker load tests

values for the response time at different request rates for a single user, for three users and the percentage of errors that occurred during the tests, respectively.

From Figure 9.1, there are distinct situations that need to be clarified. The first refers to the total time it takes to respond to simple context requests until 15 req/s. In this phase, the human context broker (HCB) performs fairly well and within acceptable real-time thresholds (2,000 ms). Still, at the current conditions, a single request takes around 620 ms to be answered (during the tests, the average round-trip time between the Apache Benchmarking machine and the HCB was 20 ms). From 15 req/s until approximately 100 req/s, the time it takes to process all requests grows linearly with the number of requests. A curious fact is that, with it, the system starts to be less stable. This is visible by looking at the standard deviation, which shows that when the load increases, so does the deviation from the mean.

Then, from 100 req/s until 200 req/s, the total time reduces drastically. The explanation to this behaviour is reflected in Figure 9.3. In short, the time decreases due to the amount of errors. In other words, if a request fails (because the HCB cannot process them or an internal error occurs), the total number of requests to be processed is inferior, and consequently, the overall time it takes to process all requests is reduced. From 200 req/s onwards, this is even more evident, as the time stabilizes around 2,400 ms. However, in this situation none of the requests are processed.

When considering a multiple request (i.e. more than one user or more than one scope), the behaviour is somehow similar but with extra delay. In this situation, the time it takes to perform a 'unique' multiple request is around 1,300 ms (i.e. two times more than a single request). Another interesting fact is that over 5 req/s, the total time to process all requests is over 2,000 ms (the acceptable threshold for real time). From this stage until 75 req/s, the total time grows linearly with the number of requests per second. Within this period, the total time to process the requests at the same request rate is about 2.5–3 times higher for a multiple request (with three users), when compared to a simple one.

Just like before, the standard deviation increases with the number of requests per second. What is interesting is that, although the percentage of errors equals 100% from 85 req/s, the system still responds to the requests until approximately 100 req/s. After that, different types of errors appear (see more details ahead). The only reason why the time it takes to process all requests is represented in the figure relates to a simple burst of simultaneous requests (and not a continuous flow of requests during 10 s – as described initially). This experiment was used to show how the system would behave under these specific conditions. Nevertheless, the values over 125 req/s have been omitted, otherwise it would not be possible to visualize what happens at lower request rates.

To better understand the previous results, Figure 9.3 depicts the percentage of errors for both types of requests and how they varied with the frequency of requests. Although the behaviour pattern is similar, multiple requests start to be affected by errors at a very early stage. Still, the types of errors encountered within the different experiments are not the same. In short, the following error types were found: failed requests, non 200 OK responses and SSL handshake failed. While the first two let the tests continue and obtain a response, the last one doesn't. However, whenever the last occurred, the previous graphics still show response times. This happens due to the fact that for these cases, a single burst at that frequency was used for the tests (e.g. Figure 9.1 at 225 req/s uses only 225 requests for the test), thus eliminating the limitation of several simultaneous SSL connections to the HCB.

A deeper analysis of the results introduced in here will be provided in section 9.6, which will compare the different outcomes of the overall CUCAF.

9.2.1.2 How do these vary when authentication and authorization are introduced?

One of the most important contributions of this work is the introduction of privacy and security mechanisms. Nevertheless, the extended benefit comes with some tradeoffs. The most obvious is efficiency. In other words, it introduces the delay to the overall communication between the different components. To better understand the implication of each procedure, this analysis will separate the authentication and authorization mechanisms.

Authentication

In short, this process consists in authenticating towards an identity provider and provide the credentials (e.g. token), within the request performed to the HE. To better understand the implications associated with these mechanisms, there are two situations that should be considered: the authentication towards the IdP and the validation of the token inside the HE. This distinction is important because the first will not affect the system performance, while the second might (the authentication can be made 'a priori').

From the literature (Sun Microsystems, 2007), it was verified that the average time for the IdP to process user credentials and generate a token (also called assertion or certificate) takes about 100 ms. Therefore, the remaining time depends on the network conditions and the size of the token. On average, this process takes around 200 ms (although this is not important). Despite the tests were realized in a controlled environment, even if a real situation takes twice the time, it will not impact the experience at all (because advertisers just authenticate once).

On the other hand, whenever a new request comes, it needs to be validated. This is done by using the token (or certificate) that comes attached with every request. Please note that unlike authorization, which may be performed on the request (or subscription) or in the answer (or notification), the authentication only takes place on the first. Although this process is simple, it involves several steps. These are described below:

1. Decoding (in this case from Base64 (Josefsson, 2006)).
2. Generate hash code (digest) for the token/certificate.
3. Decrypt the hash in the signature.
4. Compare/verify that hashes (1 and 3) match.
5. Check validity of the token/certificate.
6. Check the subject of the token/certificate.

There might be other small steps involved and this process that may slightly vary from technology to technology (e.g. verify which encoding or cryptographic algorithms were used). In this case, an X.509 certificate/token (information technology – open systems' interconnection – the directory: Public-key and attribute certificate frameworks, ITU-T recommendation) was considered. With regard to the encoding, as previously mentioned, Base64 was used. In what concerns the fingerprint/hashing function, the adopted version was SHA1 (512 bits) (Eastlake & Hansen, 2006). As for the encryption, the RSA (2048 bits) algorithm for public-key cryptography was adopted.

To evaluate the overall performance of this set of operations, the benchmarking command implemented within the OpenSSL library was used. Figure 9.4 shows the average values for each of the aforementioned steps. In total, the validation process takes on average 30.3 ms to validate the token/certificate.

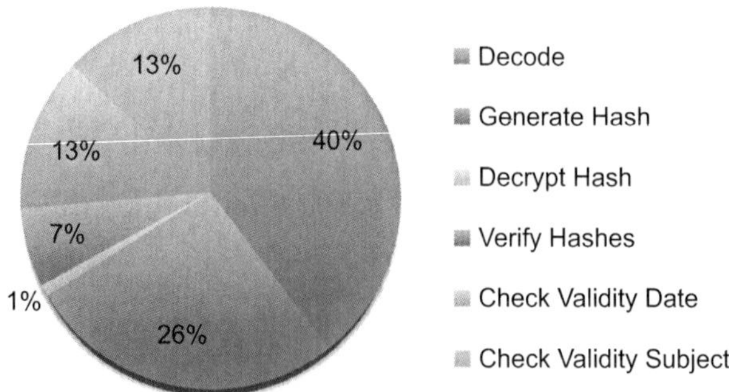

Figure 9.4 Distribution of the total time it takes to validate an authentication token

Despite the fact that these values (namely the total of 30.3 ms) represent the average for the settings of this implementation, the variation to other algorithms with different key sizes would not impact that much the overall behaviour of the system. However, authentication is only one of the steps required within the proposed approach.

Authorization

This process is more complex and involves the Policy Evaluation, Enforcement and Management Architecture (PEEM) component. Because authorization can be used at different stages, in here, this procedure will be analysed alone. This means that, independently from the authorization purpose (request, response, subscription or notification), these steps will be required. However, due to the fact that the authorization process might require extra information to perform an evaluation and/or enforcement, for now (the more complex is validated in the next subsection), only the simplest case will be covered. This means that the PEEM possesses all the necessary information to perform the evaluation and/or enforcement. For a better understanding, Figure 9.5 depicts a load test at the PEEM-1 interface for a policy containing a single rule.

This test was performed using SIPNuke (http://www.sipnuke.org) and started at 1 req/s until 500 req/s with a 15 s sleep interval between each increment (to avoid mixing the results from previous requests). In a first analysis, the average evaluation time is stable at 21 ms, until 100 req/s. Then, despite some fluctuation, there is a linear tendency to increase the time with the number of requests per second. Although the reason behind this fluctuation occurrence is not clear, from the tests performed, it seems like it is related to the way Sailfin AS manages thread allocation (to every request a new thread is created). As the experiments were performed at a local network, the delay is almost null (round-trip time is on average less than 1 ms) and consequently does not impact the previous results.

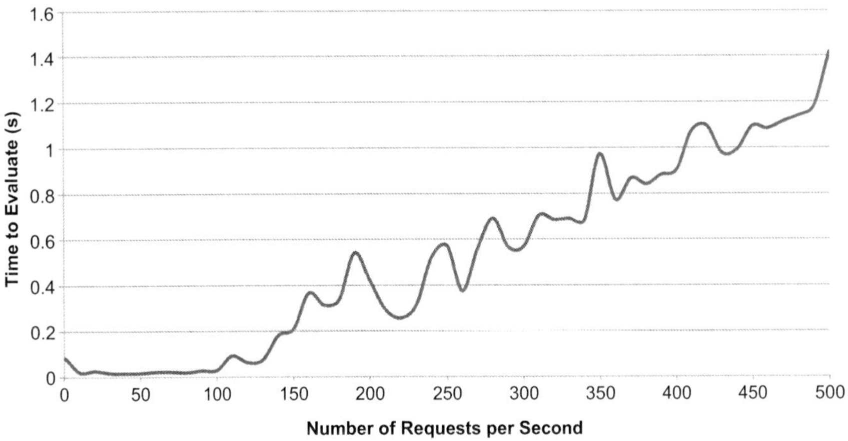

Figure 9.5 Simple policy authorization process load test (at PEEM-1 interface)

Altogether, for the simplest scenario, when introducing security and privacy into the system, an average of 51.3 ms (i.e. 30.3 ms for authentication + 21 ms for authorization) delay will be added to the overall process. To understand how these values vary for more complex situations, the following subsections demonstrate how the overall evaluation is affected by the type and number of policies and rules.

9.2.1.3 What is the implication of complex user defined policies?

As mentioned in Chapter 5, one of the innovations proposed by the CUCAF is the capability to perform authorization based on real-time context. In this sense, the PEEM component needs to collect this information (context) in real time. Therefore, during the evaluation process, it needs to get context from the HCB before accepting or denying the request. As in the previous analysis, a load test was executed. Figure 9.6 highlights how the evaluation time varies with the number of requests per second.

To allow a better interpretation of the results, it is important to mention that the HCB and the PEEM are not on the same machine or network. Therefore, alongside the measurements an average round-trip time of 20 ms was calculated. This gives a different perspective of the results, making more or less clear what is the processing time at the HCB. To simplify the experiment, and limit the amount of external factors, the context to be requested was simple (in this case 'userProfile') and available in cache. Otherwise, the HCB would have to query the respective context provider, which would increase the overall time of the operations.

The analysis of this figure is rather straightforward as it represents a mix of what was presented in Figures 9.1 and 9.5. However, in here, the load is seen in bursts and not in a continuous form. That is the reason why the overall performance is slightly better than the experiment depicted in Figure 9.1. In here, the system can evaluate the policy within 2,000 ms if the request rate is inferior to 20 req/s. The values over 10,000 ms are not depicted as the policy system (PEEM) has an internal timer of 10 s before it aborts the evaluation.

Therefore, the evaluation is practically dependent on the time it takes to perform the context lookup/request, as the contribution of the evaluation itself (without fetching the necessary data for evaluation) is negligible (i.e. 51.3 ms or 21 ms with and without authentication, respectively), when compared with the context request time (i.e. minimum of 600 ms). Please note that this task involves the following interactions: Fx1. In this sense, an optimization of this process should be considered for future work.

9.2.1.4 Does the number of existing rules affect the system?

The last experiment consisted in evaluating how would the system react when varying the complexity of a policy. To accomplish this study, the same methodology, as in a simple policy evaluation, was used. However, for this case, instead of assuming a simple policy, with a single rule, in here, the number of rules (within a single policy) varied. For the tests, policies with 1, 2, 5, 10 and 15 rules were used. Figure 9.7 depicts the aforementioned scenario.

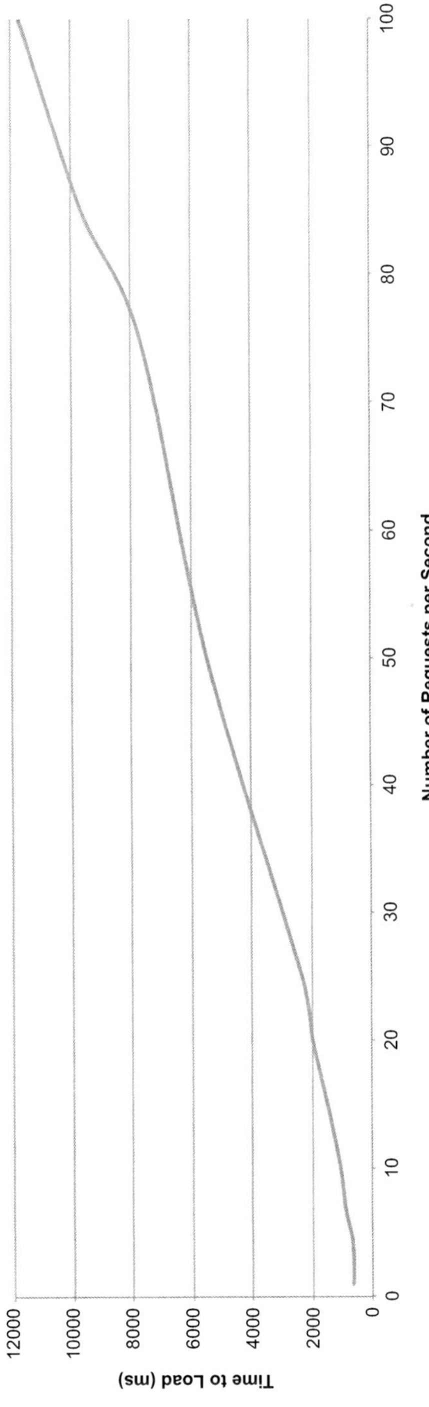

Figure 9.6 Complex policy authorization process load test (at PEEM-1 interface)

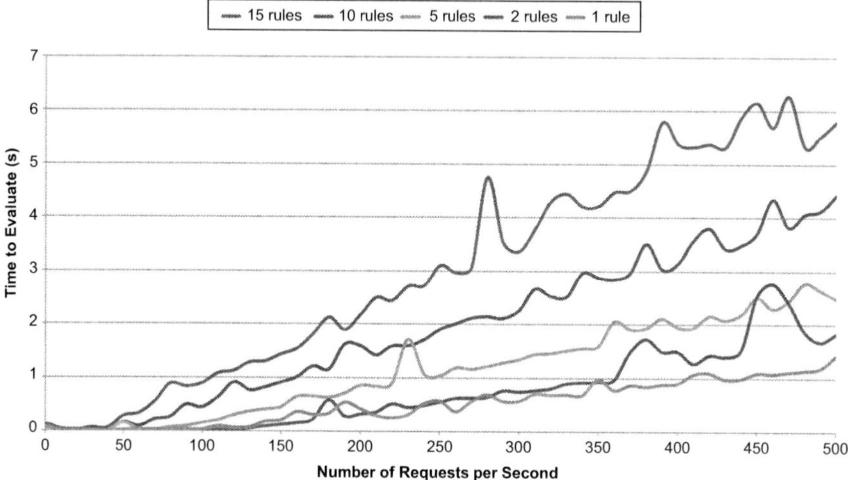

Figure 9.7 Simple authorization process load test for different number of rules

In the first instance it is easy to see a direct correlation between the number of rules in a policy and the time it takes to perform the evaluation. In fact, the correlation between each of the aforementioned experiments (i.e. 1, 2, 5, 10 and 15 req/s) is always over 0.9. From this result it is possible to infer with some degree of confidence the time it takes to evaluate a policy based on the number of requests per second and the number of rules contained within this policy. However, this is only possible because each evaluation occurs over a unique policy (otherwise this might not be truth).

Another important aspect that is worth noticing is the fact that large policies containing simple rules (i.e. that do not require extra information) perform fairly well during high-load periods. More concretely, within the proposed framework, as it was shown previously, the HCB would be the component limiting the performance of the system.

9.2.2 Evaluation of the reasoning engine

In what concerns the reasoning engine, as it is obvious, the abstraction that was proposed cannot be evaluated due to its conceptual definition. Nevertheless, the study that was performed in Chapter 6, trying to understand and predict human behaviour, can. Therefore, within this subsection, the used dataset will be described, as well as the results obtained for popularity and influence metrics. In addition, it presents the limitations found within the implementation and methodology. Likewise, an overview on how much time does each of the steps takes to be processed will be covered.

9.2.2.1 The dataset

As explained earlier, the dataset was collected from a Facebook application and stored according to the data model presented in section 6.3.2. To allow a better understanding on how information changes through time, two different snapshots are considered. The snapshots are incremental; therefore, 'Snapshot 2' contains all the information inside 'Snapshot 1', plus the extra information collected within the interval '*t*' between both. In this work, '*t*' was considered as a three-week period.

In both snapshots, the same number of users is considered. In this sense, the only values that change are relationships, interests, interactions and user's associated data. Therefore, Table 9.1 shows the total number of the overall user interests (spread over groups, event and pages). On the other hand, Table 9.2 gives an overview of the completeness of the data associated with the user personal profile.

By analysing Table 9.1, it is clear that some of the data may not be statistically relevant if some correlations are to be made. Nevertheless, that is also not the goal of this work, although some conclusions might be derived. Therefore, for data analysis (clustering) purposes, the following attributes were excluded from the evaluation: 'religion' and 'political'. To better understand what type of data was used in the experiments, the following list provide a detailed Table 9.3 resume.

Table 9.1 Total number of the overall user interests (Snapshot 2)

Users	Groups	Pages	Events
12,346	1,288	2,500	1,053

Table 9.2 User profile completeness of data (relevant data)

Attribute	Valid (no.)	Percentage
Birthday	8,432	67.8
Work	3,346	26.9
Education	5,821	46.8
Hometown	4,777	38.4
current location	3,304	26.6
Language	12,346	100
Gender	10,090	81.1
Religion	54	0.4
Political	49	0.4
Relationship status	5,055	40.6

Table 9.3 Data resume for 'Snapshot 1'

Data type	Total (no.)	Data type	Total (no.)
Album	426	Photo comment	9,756
Photo	10,780	Photo like	4,379
Post	6,025	Photo tag	3,763
Album comment	497		

Table 9.4 Data resume for 'Snapshot 2'

Data type	Total (no.)	New (no.)	New (%)
Album	445	19	4.3
Photo	11,179	399	3.6
Post	6,577	522	7.9
Album comment	509	12	2.4
Photo comment	10,017	261	2.6
Photo like	4,599	220	4.8
Photo tag	4,588	825	18.0

Snapshot 1: For a better overview over the collected data, Table 9.3 provides a resume of the entries associated to each data type for 'Snapshot 1'.

Snapshot 2: Similar to what was presented for 'Snapshot 1', Table 9.4 resumes the total amount of data considered for 'Snapshot 2'. Furthermore, it also shows the total number of different entries – New – in both cardinal and percentage values.

9.2.2.2 Limitations

Throughout the design of the overall experiments, a series of difficulties were found, which in some cases were not possible to solve. To help understanding the obtained results, in the following paragraphs, some of the limitations that were faced are described in Table 9.4.

On the prototype

Based on the advertising use-case presented, it was this work's intent to provide an example of a recommendation system based on this methodology. However, to allow such deployment, it would be required to have a list of advertisements that could be classified into the categories/clusters of interest identified by the system. If existing, the recommendation could be made using the methodology proposed in Crandall et al. (2008), where the modelling of activities (what to advertise) and interactions (who and when to advertise) is highly dependent on correlated interests/preferences, influence and past activities/interactions of both the user and their friends (or influencing peers). Another limitation is that a feedback channel is not available to measure the accuracy of the interest's correlation predictions, as well as the reach of the cascades. Either way, this is an open door for future work.

On the dataset

Despite one of the main strengths of this work is based on its detailed dataset, it is also here where most of the limitations arise. Due to the sensitivity of the data being collected, Facebook is forced to protect the privacy of their users. In this sense, when crawling data, there is no way for the system to know, which of the users granted permission to each of the attributes being requested. In addition, most of the data collected is based on a small set of active users (the ones that actually used the application), representing approximately 0.4% of the total number of users. In addition, the amount of data being collected also depends on the privacy settings between the users themselves. This limitation raises a concern with the data reliability, as the relationships and interactions may not be as accurate as they should (e.g. user A might not be able to access the entire collection of photo comments from user B). As a consequence, the level of granularity in the information may vary from user to user. As an example, assuming the crawl is made from user A account, the system can check which friends A has in common with user B, but it is not allowed to get the entire list of user B friend relationships. The same principle applies to other types of data (e.g. other relationships and interactions). This may lead the results to be biased towards the users that installed the application.

As seen in section 6.3.3, some of the data related to the user personal profile is missing. This might be because of privacy issues, or simply because users don't want to share this information. Even worst, as the users themselves type most of these fields, they may contain misspellings, might be written in different languages or filled improperly on purpose. As a consequence, if no correlation (or proper clusters) can be found, these attributes are excluded from the multi-feature (dimension) clustering process. However, the biggest limitation arises from the Facebook platform itself. Due to the amount of friends, relationships and interactions that some users possess, together with the complexity of the queries performed by this application, in some cases, the application can only obtain a sub-set of the results. Moreover, by analysing the data in snapshots, the results become dependent on the time window between snapshots. A short window might not be enough to capture a cascade event, while a long one might obfuscate some evolutions (cascades) due to other factors. Another issue relates to the fact that it might not capture the ordinarily of data entries, and therefore miscalculate the predictions. Nevertheless, these problems can be minimized due to successive crawls, which are promptly time stamped by the system. This is only possible because the users allowed the application to make offline queries about their data.

On the processing power and memory

While performing consensus clustering, there was a limitation in terms of processing power and memory. While the first would only delay the clustering operation, the memory outage could not be fixed. Therefore, as reducing the number of users would change the expected results, it was decided to perform two separate consensus clustering operations. In this sense, features related with interests (e.g. groups, pages, events) and features related to demographics (e.g. age, gender, location) were clustered independently. Still, the results were limited to five clusters per type of features.

9.2.2.3 Evaluation techniques

With no public benchmark available, it is not possible to compare the obtained results with other works in this area. Therefore a 'gold standard' was created within our work. The following subsections show the methodologies used for both one- and multi-dimension clustering.

Pooling

In order to estimate the accuracy of the obtained clusters, a well-known pooling technique (Aslam et al., 2003) was used. Consequently, the benchmark consisted the following operations:

1. Randomly select 10% of clusters from a result.
2. From these 10%, if cluster has less than 10 elements, evaluate all of them. Otherwise, randomly select 10 elements plus 10% of the cluster and evaluate it.

The evaluation consists in human monitoring the obtained results (cluster structures). More concretely, it consists in assigning two types of labels: 'true' (1) and 'false' (0), which correspond to 'it's a right answer' and 'it's not a right answer', respectively. Then, the accuracy is used as a metric to decide whether the cluster structure is valid or should be discarded from further analysis. Its value is given in (9.1).

$$\text{Accuracy} = \frac{\text{Num}_{\text{True}}}{\text{Num}_{\text{True}} + \text{Num}_{\text{False}}} \tag{9.1}$$

Therefore, when the accuracy value is above a determined threshold 'gold standard', the cluster is considered as valuable or valid.

Davies and Bouldin's validity index

Consensus clustering validity can be measured in several ways:

- Use classical validity indexes on data representation in the original space.
- Use validity indices and rely on the agreement between the consensus partition.
- Use modified validity indexes on similarity space.
- Use statistical indexes based on pair-wise similarity.

In this case, the first methodology cannot be used because there are different original spaces for each single-feature cluster structure. Therefore, there is no common original space. Regarding the statistical methodology, it is too expensive due to the enormous dataset size. In this sense, we propose the use of Davies and Bouldin's validity index (Davies & Bouldin, 1979) on similarity space, in order to measure multi-clustering validity. This index is easy to calculate and it produces good results in consensus clustering evaluation (Duarte et al., 2010).

9.2.2.4 One-dimension clustering

One of the most important procedures for clustering is tuning the input parameters for consensus clustering. Because each feature contains different types of data, this task must be performed to every single feature. As a consequence, these parameters will vary the accuracy of each cluster structure. To better understand the obtained results, Table 9.5 shows the distance function and respective parameters used, the total number of clusters, as well as their accuracy for every single feature

Table 9.5 Resume of one-dimension clustering process (Snapshot 2)

Dimension	Algorithm	Epsilon	minPoints	Clusters	Accuracy (%)
Age	Euclidian	0.1	5	20	100
Gender	Euclidian	0.1	3	2	100
Language	Euclidian	0.1	7	13	100
City	Manhattan	0.5	15	29	100
Country	Manhattan	0.4	10	26	100
Work	Manhattan	0.1	5	29	100
Education	Manhattan	0.3	44	19	90.3
Relationship	Euclidian	0.1	9	9	100
Group	Manhathan	0.5	3	42	76.5
Page	Manhattan	0.5	3	65	68.7
Event	Manhattan	0.5	3	41	59.4

(dimension of data). These represent the best combination of the input parameters for each feature. The accuracy of the results was calculated using the methodology described in section 'Pooling'.

From the following results it is possible to see that some features are more stable than others. Another important aspect is that the number of clusters depends on the initial parameters, and these were sized to prevent too big or small cluster structures, as the former would not allow further consensus clustering, while the latter would limit the final multi-dimension clustering results.

9.2.2.5 Multi-dimension clustering

As mentioned before, due to computational and memory limitations, the analysis of multi-dimension clustering is performed in two separate classes. The first refers to user demographics, while the second to their interests. In here, Snapshot 2 was used as data source for the presented results.

Concerning demographics, it was decided to use age and relationship status. As demonstrated in Table 9.5, the total numbers of clusters were 20 and 9, respectively, covering a total of 9,899 users. Regarding interests, the used features were groups, pages and events, containing 42, 65 and 41 clusters, respectively, resulting in a total of 9,006 users. The parameters used for consensus clustering were: no sampling and Self-Organizing Map (SOM) algorithm with 200 iterations. Table 9.6 resumes the obtained results.

Table 9.6 Resume of the multi-dimension clustering process

Demographics	Features	Users	Interests	Features	Users
Cluster 1	4	3,137	Cluster 1	2	3,563
Cluster 2	18	941	Cluster 2	2	2,115
Cluster 3	4	1,797	Cluster 3	73	101
Cluster 4	15	3,401	Cluster 4	70	182
Cluster 5	23	623	Cluster 5	149	3,045

Due to the limitation of five clusters per type of data, it makes no sense to validate these results, as it is known beforehand that these would not be satisfactory. However, based on the method described in section 'Davies and Bouldin's validity index', it would be possible to assert it. Despite the limitations, it is possible to verify some interesting facts. Clusters containing a huge number of users usually have fewer features associated with it. These are the less interesting ones, because they do not provide much information about the users. Moreover, the few they provide is very generic and therefore less interesting for the use-case (advertising). On the other hand, clusters with a smaller amount of users are generally associated with more features, meaning that it is possible to know more about these users, in a more personalized way. In addition, it is possible to see what kind of features (coming from different dimensions) are normally seen together. This can be very interesting for advertisers, as they can, not only better contextualize their ads, but also understand their customer's behaviour.

As future work, it is necessary to include all the available features in the clustering process (especially for demographics), to improve the accuracy of the results. Moreover, to overcome the computational limitations, the proposed methodology should be explored under a cloud (or clustered) computing environment such as Amazon E3C4[2] or Heroku.[3]

9.2.2.6 Popularity

As explained in section 6.3.3.3, popularity is calculated within each cluster to understand which are the most 'important' users. Although the method can be applied to one-dimension clusters, in this work it was decided to evaluate the user's impact at a multi-dimension level. To simplify the interpretation of the results, the analysis of popularity was separated in user demographics and interests. Figures 9.8 and 9.9 represent the aforementioned, respectively.

For a better representation, all users with a popularity value of 0 were excluded from these graphs. Moreover, for both cases, Snapshots 1 and 2 are presented to facilitate the study. From both figures, it is possible to verify that independently from their source it is possible to identify some users with a significant higher value of popularity. These are called leaders.

In order to identify relationships between the data in different snapshots, it is necessary to find a match between clusters over time. These were identified by the maximum number of users in common between both samples. This means that Cluster 1 in Snapshot 1 does not necessarily correspond to Cluster 1 in Snapshot 2. Despite the sample is not optimal, the average correlation found between both snapshots was 0.79 for demographics and 0.80 for interests. Furthermore, although not visible here, the leaders are likely to maintain over time, that is, leaders of a cluster in Snapshot 1 are more or less the same as the ones in Snapshot 2.

[2] http://aws.amazon.com/ec2

[3] http://heroku.com

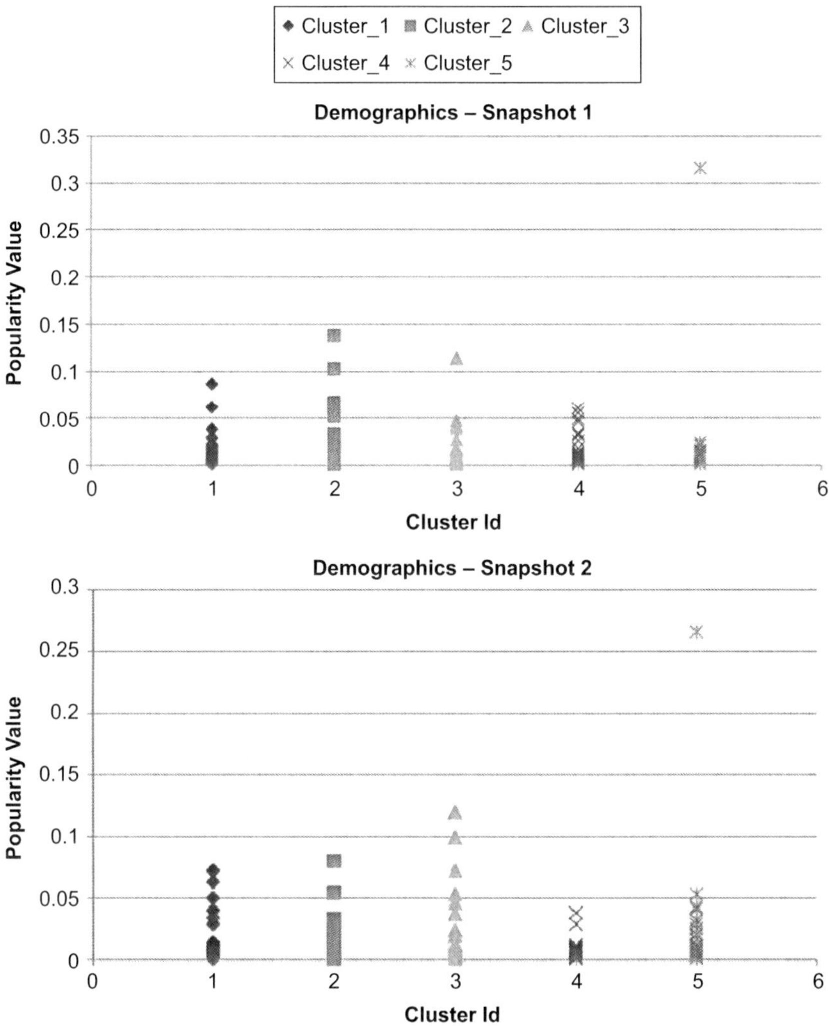

Figure 9.8 Popularity values for demographics in Snapshots 1 and 2

Despite there is not enough data to make assertions, this trend was verified in the results. Depending on the application, the number of users considered as leaders may vary. For these experiments, a popularity value of 0.002 was considered as a threshold, meaning that all users with values above this number are considered as leaders. Based on these data, it is possible to understand, which peers are more 'important' within a specific group of interests or demographics.

In the future, these results should be compared with other strategies to calculate popularity, namely based on known network analysis metrics, such as centralities (e.g. degree, betweeness, closeness, etc.), which evaluate relationships between all the peers belonging to one cluster.

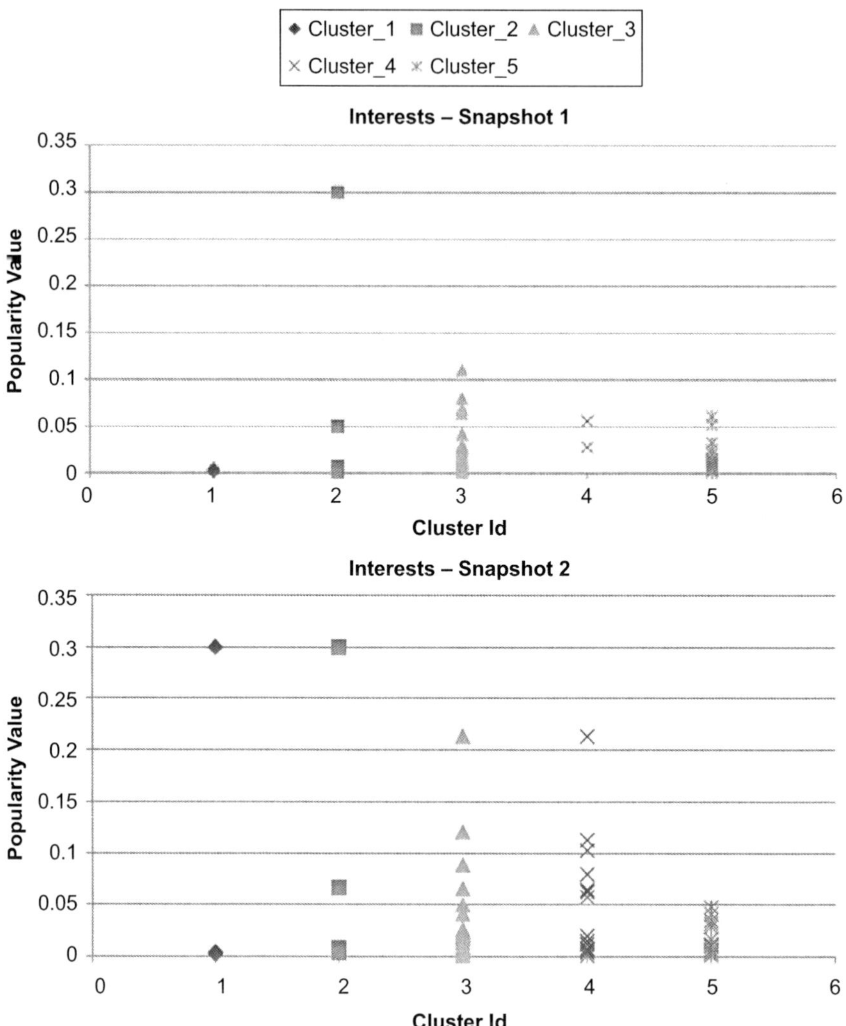

Figure 9.9 Popularity values for interests in Snapshots 1 and 2

9.2.2.7 Influence

As mentioned in section 6.3.3, the evaluation of influence will only focus on the analysis of new users joining a previously known cluster. In short, the strategy consists in verifying if the number of recent interactions (between the two snapshots) is higher among popular users than regular users inside the same cluster. Tables 9.7 and 9.8 present some examples of influence calculation for new users in demographics and interests cluster structures, respectively. Moreover, they include the average value of influence whenever it was possible to calculate. For some users, the influence was not possible to calculate because there were no interactions with other peers belonging to the cluster.

Table 9.7 Sample of influence calculation for demographics

User	Number of I_{t1} with P_u	Number of I_{t2} with P_u	Total interactions with all users	Influence
A	2	3	20	0.15
B	16	18	18	1
C	39	40	1645	0.0243
D	706	707	762	0.9278
E	100	115	1645	0.0699
			Average	0.06465

Table 9.8 Sample of influence calculation for interests

User	Number of I_{t1} with P_u	Num. of I_{t2} with Pu	Total interactions with all users	Influence
F	7	9	19	0.4737
G	5	7	27	0.2593
H	0	1	60	0.0167
I	12	13	482	0.2691
J	78	79	135	0.5851
			Average	0.2306

Despite the results show some indication that there is a relationship (of dependence) between new users and their interaction with previously popular users, this correlation is not so high. Still, this trend is particularly noticeable for demographics. Nevertheless, the fact that these values only account users where some sort of interaction occurred cannot be ignored. If not considering this situation, the average would be even lower. There are two main reasons for this behaviour. The first relates to the initial limitation of five clusters, which obviously does not gather the necessary granularity level. As a consequence, clusters are very big and the number of interactions among their peers is also reduced. The other relates to the amount of features that were used for clustering. It is obvious that age and relationship status alone aren't good indicators of influence. Nevertheless, when adding location, education, work or language, these results might be completely different (more targeted). Therefore, in the future, by taking advantage of cloud computing or any environment without computational limitations, it will be possible to explore further combinations of features and 'play' with a higher number of clusters.

9.2.2.8 Step-by-step processing time

As depicted in Figure 6.2, there are different steps associated with the calculation of popularity and influence. These are:

- Extract information from the social network (Facebook).
- Perform one-dimension (single-feature) clustering.

- Perform multi-dimensional clustering.
- Calculate popularity.
- Calculate influence.

This section will analyse how long it takes to perform each of the aforementioned tasks. The importance of this subject relates to the system ability to address real-time scenarios or not.

Extract information from Facebook

Although the time it takes to fetch all the necessary information from Facebook regarding a specific user cannot be generalized, from a previous research initiative (Simoes et al., 2010a), it was verified that the number of friends can be a good predictor. To verify the previous assumption, Figure 9.10 represents the time it took to load all user-related data (profile, comments, likes, photos, groups, events, pages and relationships) for every single user that installed the 'social enabler' application.

Although the previous principle is verified, there are a few other interesting things that should be noticed. There are three distinct phases that can be observable in this process. To better understand these, a polynomial fit of third degree was sketched on the figure. On a first stage, independent from the number of friends, the time it takes to load all user-related data is more or less the same and very low (below 30 s). However, this is only valid until approximately 100 friends. Then, until 750 friends there is a linear correlation of 0.6 (see dotted line in Figure 9.10). In other words, it is clear that for this interval of users, there is a relation between the number of users and the time it takes to load their information. For the last group of users (with more than 750 friends), it is possible to see that the time it takes to load their data is more or less constant and stable.

The interpretation of such results is explained by the limitations presented earlier. While it is easy to understand that users with fewer friends are (usually) expected to interact less (consequently less data), it should be expectable that the ones with a large amount of friends would behave in the opposite way. Nevertheless, there are two main reasons that prevent that from happening. The first

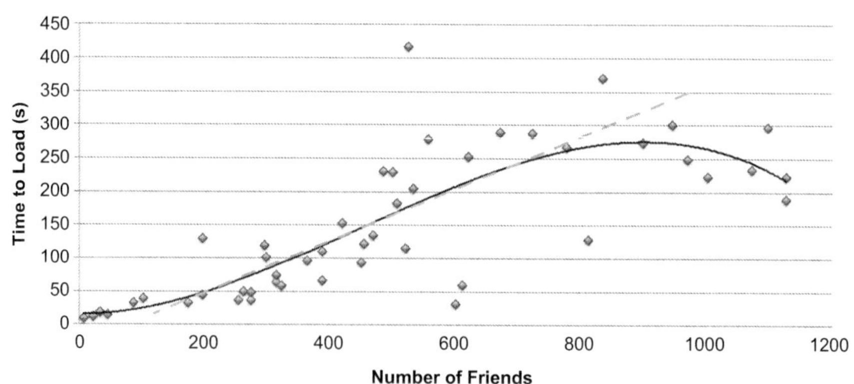

Figure 9.10 Time to load all user-related data vs. number of friends

derives from social studies, where it has been proposed that a person cannot keep active friendship relationships over a determined threshold. This value is often referred to as the Dunbar limit (1992). Consequently, no matter how many friends a person has, it is not expected that the interactions grow proportionally or exponentially with it. The second reason relates to the Facebook query limitation. Whenever a user has more than a pre-determined number of photos, comments or other associations, Facebook only returns a sub-set of them. In fact, in a first instance, it exits with an error and the application is then responsible for querying a smaller sub-set. The group of users whose number of friends varies from 100 to 750 follows a rather regular pattern.

At first sight, these results many seem irrelevant; nevertheless, this simple verification can be applicable to many other areas. Knowing that the number of friends (connections) is also a good estimate for the amount of data available about a particular user, it can be used for network dimensioning, load balancing, data storage strategies, query mechanisms, etc. Altogether, understanding user's usage of a system is the key to any service or product success.

Perform one-dimension (single-feature) clustering

While the collection of user-related data is done separately for each user, the remaining steps are applied to the overall dataset. Before getting into the analysis of one-dimension clustering operations, it is important to make a reference to the pre-processing methods. These are executed by simple SQL commands, which are responsible for cleaning data (remove spaces, invalid characters, etc.) for execution. The total time to execute all these operations for a dataset containing 12,346 users is less than 5 s and therefore can be considered negligible when compared with the following steps.

As mentioned before, for each type of data (dimension or feature) a cluster structure was generated. To better understand which parameters would provide more accurate results, during the experiments, several trials were performed, where the input parameters 'epsilon' and 'minPoints' were alternated. During the tests, it was possible to verify that the time it takes to perform the calculations varies drastically with the inputted parameters. Moreover, it was also seen that most of the times the clustering operation was very inefficient. Therefore, this analysis will focus on the parameters detailed in Table 9.5 (which provide the best results). To help with the observations, Table 9.9 resumes the time it takes to perform the clustering operation to every single dimension. Furthermore, it includes the number of users that fit the criteria (after data cleaning), as well as the number of elements being grouped per dimension. To complement this analysis, Figure 9.11 shows the distribution of the total time necessary to perform all one-dimension clustering activities, by feature.

At first sight, a couple of observations can be made. The time it takes to perform clustering is somehow related with the amount of data involved (which is obvious). However, there is no clear indication on how this relationship is established. Going a bit deeper into the analysis, it is possible to see that the length of elements influences the time of the analysis. As an example, the time it takes to

Table 9.9 Time necessary to perform clustering for each dimension

Dimension	Time (s)	Users	Elements
Age	11.45	3,137	48
Gender	106.6	10,090	2
Language	2.8	12,346	36
City	12.44	4,777	933
Country	19.12	3,447	670
Work	176.46	3,453	2,737
Education	304.68	5,974	4,909
Relationship	797.5	5,179	9
Group	3,019.7	10,020	1,305
Page	3,901.3	11,203	2,647
Event	2,132.5	8914	1091
Total	10,484.55		

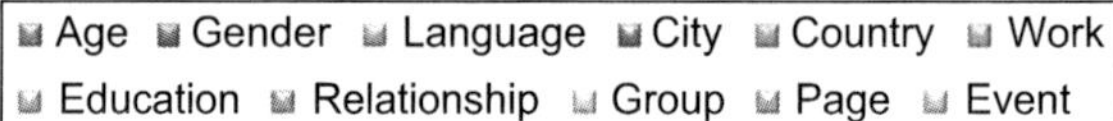

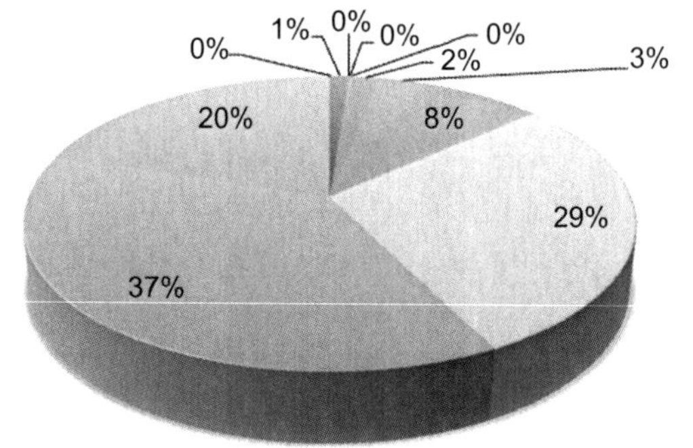

Figure 9.11 Distribution of the time necessary to perform one-dimension clustering for all features

cluster 'language' (stored in two letters, e.g. pt, de, uk) is much less than 'gender' (stored as 'female' or 'male'), even though the number of elements is lower in the first. Therefore, the more complex and long the elements are, the longer this operation will take. When comparing these results with the ones in Table 9.5, it is also not possible to correlate the number of clusters with the amount of time to generate them.

Still, there are some lessons that can be taken from this analysis. First, understanding how long it takes to process each feature can help planning the deployment of applications or processes that depend on this data. Second, when run in parallel, the operations do not necessarily need the same amount of time (it is usually faster). But probably the most important lesson relates to the way data is managed throughout the entire data mining process. It is clear that some optimizations could have been made. A simple example would be assigning numbers to strings (do a mapping) to improve the speed of clustering. However, this can only be done when the strings themselves are not being used for comparison. For example, if one tries to find, which groups cluster together based on their name, this technique would not work. However, if one tried to compare, which groups cluster together, based on the set of common users, then it would make sense. Finally, this study can also be used to compare with other types of data or algorithms.

Perform multi-dimensional clustering

Despite the most interesting results regarding the consensus clustering technique have been shown, it is important to understand how these were generated and whether they could be used for real-time activities. As opposed to one-dimension clustering, changing parameters did not have any impact on the time it took to perform the operations (only on the accuracy). However, there are three parameters that influence it directly. The first and the most obvious is the number of iterations. As the process is repeated, the dependency on this variable is linear, that is, an operation with 200 iterations will take 20 times more than the equivalent with 10 iterations only. The second factor relates to the number of users being considered. To better understand how it is influenced, Figure 9.12 depicts the relationship between the number of users and the average time it takes to perform clustering with 148 features resulting in five clusters (based on 15 trials).

From this representation (Figure 9.12) it is possible to understand that, although there is not an exponential growth, it is also not linear. Inn the figure, in light dark, there is also a curve fitting the data. It is represented by a 2nd order polynomial. Please notice that features are described in a binary format (0 or 1), and consequently do not matter if they relate to demographics or interests.

Looking at the figure, from a certain amount of users (approximately 2,000, which take more less 8 min), the clustering operations may not be suitable for real-time purposes. However, in here, real time can be considered in another time perspective. While for multimedia applications real time is in the order of milliseconds (and most of the time below 1 s or 2 s), when it concerns to data mining, this

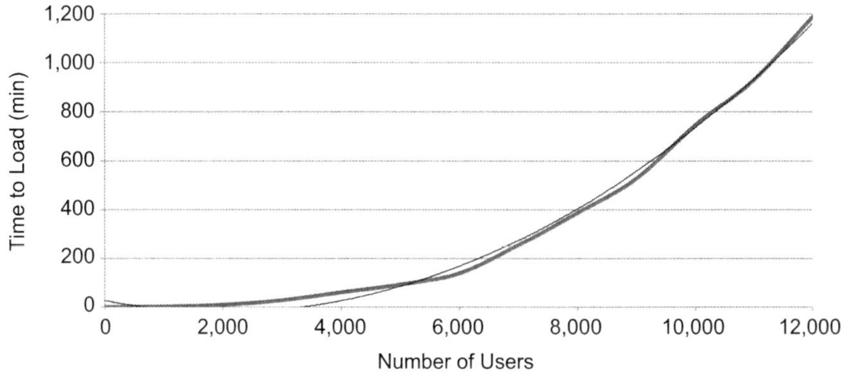

Figure 9.12 Number of users vs. time necessary for clustering 148 features

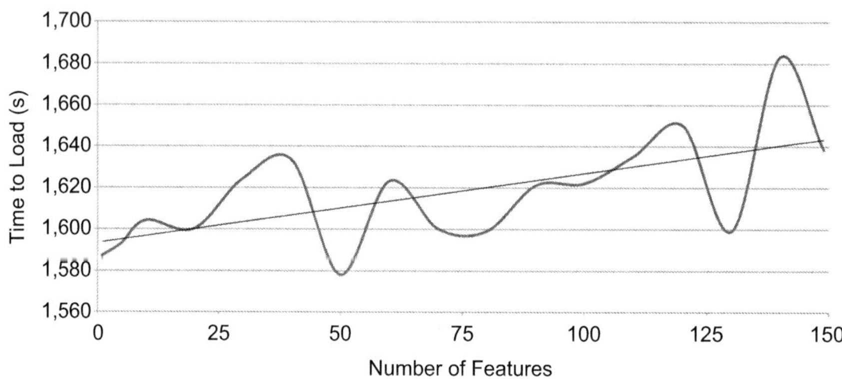

Figure 9.13 Number of features vs. time necessary for clustering 3,000 users

threshold might be different. Moreover, it should be pointed out that this performance is highly correlated with the hardware being used for the experiments.

The final parameter that can influence this process is the number of features. To study how this parameter affects the total time, Figure 9.13 shows how the time it takes to perform clustering varies (in average) with the number of features for 3,000 users (based on 15 trials).

In a first instance, as highlighted by the light dark fitting line, it is possible to see a slight relationship between the number of features and the time it takes to perform the clustering operations. However, this is only possible due to the scaling in the graphic. In reality, when comparing both Figures 9.12 and 9.13, it is possible to see that they are represented in different scales (minutes and seconds, respectively). Therefore, making a more precise analysis, it is possible to notice that for the amount of features considered (between 1 and 150), the time it takes to cluster the entire matrix is not that much affected. Altogether the variation between the maximum and the minimum is less than 120 s.

The explanation for the difference in the behaviour between adding features and users is rather simple. In short, adding a user represents another row in the consensus matrix, while adding a feature a new column. In terms of processing, they both represent the same load; however, in any kind of real system, the number of features is expected to be a rather stable field, while the number of users will grow with time. Seeing from another perspective, adding 150 features is more or less equivalent to adding 150 users, and this is why the impact is not so high.

Still, every system has a limit. Therefore, due to the memory limitations in the system used in this work, the number of features had to be reduced (divided into demographics and interests), not because of its high impact, but due to the amount of users being considered (i.e. 12,346 users).

Calculate popularity and influence

In what concerns the calculation of popularity and influence, the time it takes to calculate them is directly related with the processing power of the machine.

Basically, the formulas were implemented in Java and a small script was designed to test different cluster structures (with different sizes and number of relationships). Due to the fact that the statistical value of any interaction, such as a shared photo, a common event or a simple comment, counts as an operation, their contribution towards the final result is the same.

Because these results alone are not very interesting, in here, only the project results are presented. In this sense, considering the 12,346 users before multi-feature clustering, and assuming the 10 clusters generated (i.e. five demographics + five interests), the total number of users in all cluster structures was 18,905. Concerning popularity, it took less than 50 s to calculate the popularity value for each user within its cluster. Regarding influence, as this study only considered one of the methodologies (see (6.2)), the total time was less than 20 s. Please notice that this process also accounted the matching between clusters, which is a mandatory step before calculating influence.

Altogether, the overall process totals around 1 min and 10 s. Obviously, these values will scale proportionally with an increment of users, their relationships and also the number of clusters (due to memory limitation we could only generate 5, which reduces the granularity of the operations).

9.2.3 *Evaluation of the session management enabler*

In what concerns multimedia delivery, there are several aspects that need to be evaluated. Although the number of features the SME allows is not important on this part, the test scenarios identified in section 7.5 need to be validated by experiments. In this sense, based on session initiation and modification use-cases, the following issues are analyzed:

- What is the impact of many parallel sessions and sessions with many users?
- What is the overall system delay and how is it distributed across components?

Just like in the previous subsections, whenever possible, the tests will be performed in isolated components to understand the exact limitations of each. After these experiments are concluded, an end-to-end evaluation will be provided. For the following tests, except when explicitly indicated, to avoid delays associated with monitoring and measurements, all values were captured using Wireshark.[4]

9.2.3.1 What is the impact of many parallel sessions and sessions with many users?

To understand how the system reacts under different stress conditions, two different tests are proposed. They both are based on the session initiation use-case (presented in section 7.5.1) and have common assumptions. The first relates with the communication with other entities. To eliminate external factors, all users and all sessions are associated to the same content identifier (no communication with the Content Processor and Delivery (CtPD)). Likewise, the user context considered is

[4] http://www.wireshark.org

the same for every participant (no communication with the HCB). In addition, the network & transport layers were emulated by an internal function with pre-defined sub-groups for each user. This implies that the overlay trees are replaced by different multicast addresses at the local network. For these tests, two audio and two video sub-groups were used (four multicast addresses).

Taken into account that both the clients and the multimedia delivery gateway (also known as CtPD) were emulated by the SIPNuke tool, the unicast scenario was not tested as it would imply many modifications on the code and would not bring any advantage, as these tests do not account multimedia delivery. To facilitate the understanding of the results, three different metrics are defined. The first is D_{pro}, which denotes the time between the service invocation and the first Session Initiation Protocol (SIP) message sent towards the client, or the CtPD. The second is D_{sig} and it encompasses the time between the first SIP message issued at the SME, and the last SIP message (200 OK) associated with that session, received from the client or the CtPD. Finally, the D_{total} represents the sum of D_{pro} with D_{sig}.

Varying the number of sessions

To perform this experiment, each group/session consisted of five users (to make it minimally realistic). The service (or session initiation) call rate was tested with 1, 5 and 10 req/s. To make sure that the total number of parallel sessions was cumulative, the session release (for all requests) occurred only 20 s after the last request. Consequently, the respective SIP BYE messages were sent to free the resources for the following batch of tests. The final results are depicted in Figure 9.14 and represent the average values obtained by running the script for 20 times. Before getting into the analysis, it is worth mentioning that the tests were performed in a fast-Ethernet LAN (100Base-T).

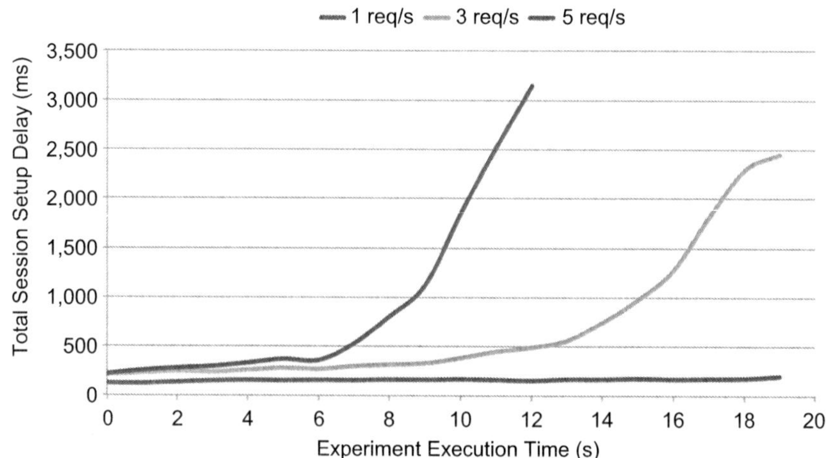

Figure 9.14 Session setup delay (D_{total}) for different session initiation call rates

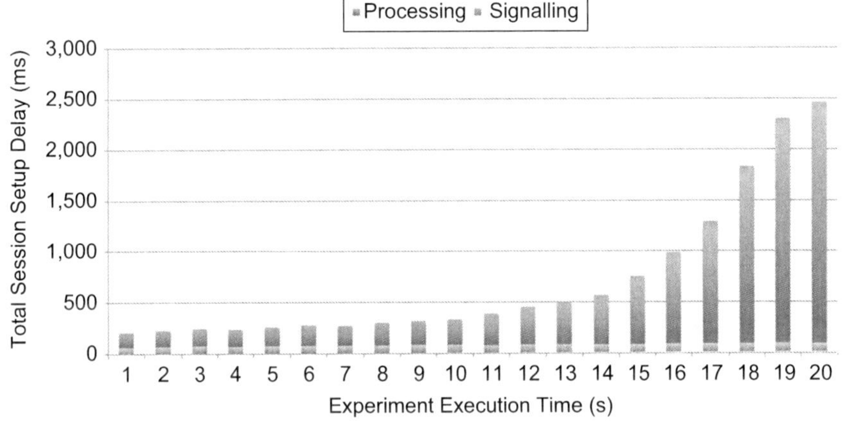

Figure 9.15 Distribution of D_{total} through D_{pro} and D_{sig} for 3 req/s

As it was expected, the performance of the SME is affected by the number of concurrent sessions or users involved. However, this correlation is not straightforward to interpret. As it is possible to observe, independently from the scenario, until approximately 30 concurrent sessions (the number of sessions is calculated through the multiplication of the experiment time elapsed, multiplied by the respective call rate), the system behaves in a very stable way, less than 360 ms. It is important to notice that this is equivalent to 150 (30 sessions $\times$ 5 users per session) users in parallel. Above this value (30 concurrent sessions), the total time it takes to establish a session starts to increase almost exponentially. To help us understand what causes such behaviour, Figure 9.15 shows the distribution of the time taken for processing and signalling at a 3 req/s rate (could be any other rate).

When analysing Figure 9.15 it is easy to understand what is the factor behind the increase in the total delay – processing. Nevertheless, the processing time can be derived from several instances. As mentioned, most interfaces that should be involved in the processing have been emulated, and therefore these should not explain these results. In this sense, the most plausible explanation refers to the limitations within the Sailfin application server, which is not capable of efficiently managing too many parallel sessions under heavy call rates. This is partially explained by the fact that a new thread is created by each session that is initiated. Again, it should be pointed out that each session on the SME web service side (invocation of session initiation) corresponds to a total of six sessions distributed by the clients (5) and CtPD (1).

In what concerns the signalling delay, the main reason why it remains practically constant is the fairly low call rate (although a similar behaviour is noted for 1 and 5 req/s). The load on the networks and respective IP Multimedia Subsystem (IMS) network is not representative. However, this scenario could be different if multimedia delivery was involved. If the network was highly loaded with Real-Time Transport Protocol (RTP) packets, it would be expectable that

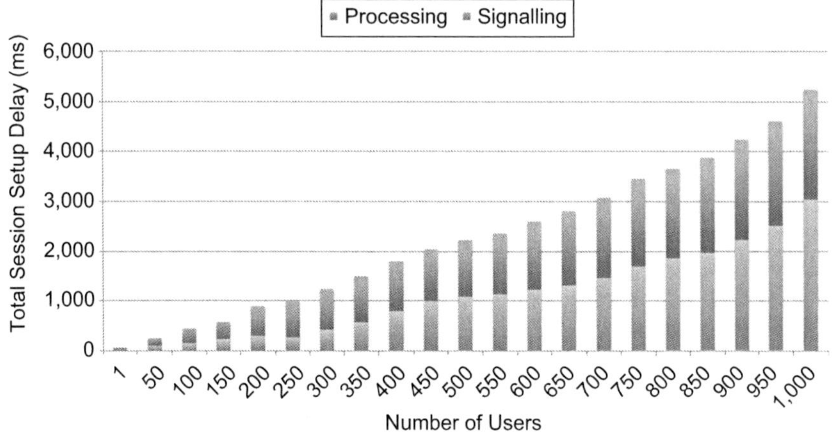

Figure 9.16 Session setup delay (D_{total}) for a single session with different number of users

retransmissions could occur, making the signalling delay increase (in some cases exponentially, with the number of users and requests per second).

Varying the number of users

Previously, it was shown that increasing the number of sessions and consequently the number of users, the total delay to set up a session would increase. In here, by considering a single session, several tests will be done, where the only parameter that changes is the total number of users per session. The tests were performed from 1 until 1,000 users, with increments of 50 users. Figure 9.16 introduces the results for this experiment.

To avoid overloading the client side, 10 different instances of SIPNuke were running, emulating 100 different clients each. In this scenario, it is possible to see that the delay increases with the number of users. Still, the SME is capable of establishing a session for 250 users in less than 1 s, which is extremely interesting. This is only possible because the tests were realized in a local network and the user contextualization was set to a minimum (i.e. fixed context and sub-grouping). With these conditions, the contribution of the time spent in signalling and processing is more or less the same. However, to be rigorous, when the number of users involved in a session is higher than 800, the time associated with signalling becomes higher than the processing one. This is probably related to the fact that some retransmissions might occur (although it also depends on the server configuration). Just like in the previous scenario, if media delivery would have been involved, the results could be totally different (the delivery itself is outside the scope of this work). Nevertheless, to overcome these analysis simplifications and limitations, the following subsections will cover end-to-end scenarios, with real components and not emulators or fixed values. However, due to the difficulty of provisioning users on the HCB, the number of users involved in the upcoming analysis will be reduced.

9.2.3.2 What is the overall system delay and how is it distributed across components?

The previous subsection showed how the system reacts to different loads of users and sessions in parallel. In here, the overall system delay will be evaluated by using the session initiation, session modification and context change use-cases (add user and change context). In each of them, a distinction will be made between the time spent between the different component interfaces (i.e. communicating) and the time spent processing in each of the entities.

For these tests, a single session with five users was used. Furthermore, a single user was added, and the context change refers to the noise notification, which triggers the change of the session at the user side. A concise description of the communications and interfaces between the different components is available in sections 7.5.1, 7.5.2 and 7.5.4. Just like in previous experiments, the clients were emulated by SipNuke instances. All the results presented represent an average of consecutive testing performed in different times of the day, during an entire week. This helped removing some 'noise' that could be inserted due to particular constraints at the time of testing (network or processing issues). This is particularly important as some of the adjacent components (namely the HCB) could be used simultaneously by other applications. To better understand the behaviour, Table 9.10 introduces the measured results for all scenarios, while the following subsections detail the analysis for each of them.

In a first overview of the results, it is possible to see that session setup is a bit slow when compared to the other use-cases. Another important aspect concerning these tests relates to the fact that authentication and authorization mechanisms were not active in this stage (as it would introduce extra delays and it would be hard to monitor).

Session initiation use-case

As it was shown previously, the average amount of time it takes to initiate a session without any kind of context for five users is around 155 ms. In contrast, when context is added, this value goes up to 4,332 ms. To better understand how the extra delay is introduced, Figure 9.17 represents the relationship between the time spent in the different stages of the initiation scenario.

Table 9.10 Time (in ms) spent per interface/component for different test scenarios

Interface and component	Session initiation	Session modification	Context change
SME – CtPD	666	–	–
SME – HCB	2,634	623	963
SME – NTL	29	24	28
SME signalling	54	5	6
SME processing	949	114	122
Total	4,332	766	1,119

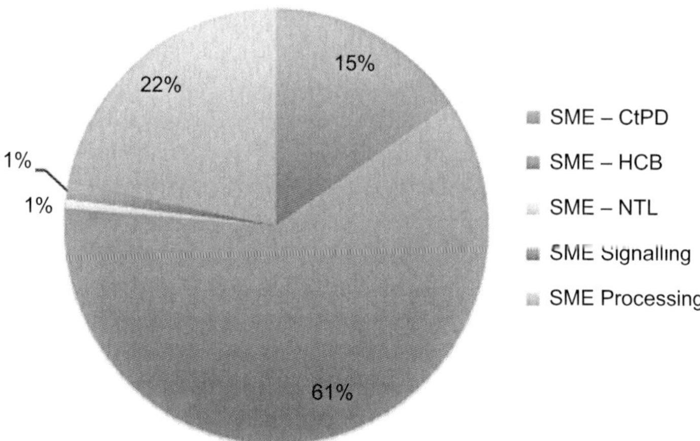

Figure 9.17 Comparison of time spent on session initiation for five users within a single context-aware session

From the Figure 9.17 it is possible to see that most of the time (61%) is spent on requesting context. Therefore, a more efficient way should be found to request user context more efficiently. Either way, as the triggering occurs at the server side, the users are not aware of such delay. In here, a communication towards the NTL is considered because its delay time is very less (29 ms). Then, the communication with the CtPD takes around 15% of the time. For this specific use-case, the number of different encodings related to the multimedia item was 6 (i.e. 4 video and 2 audio). In what concerns the processing time, it is explained due to the parsing and matching of context originated from the HCB and the CtPD, plus the time it takes to process the SIP sessions on the server side. In here signalling is negligible because the tests considered emulated clients running on the local network. However, in a parallel test, with an Android client connected through a wireless network, the average signalling delay went up to 1,400 ms. This means that in a real-case scenario, the session initiation use-case would actually take around 5,700 ms to be completely realized.

Session modification use-case

Whenever a new user joins a current session, the process is similar to the session initiation, although it involves a single user. The exception lies in the absence of communication with the CtPD (unless it is necessary to start a new sub-group). For this use-case, it is assumed that all the sub-groups were previously initiated. Based on these conditions, Figure 9.18 represents the relationship between the amounts of time spent in different stages of the modification (add new user) scenario.

In this use-case, as previously, the time spent on the user context acquisition is the most representative (81%). However, this process is not critical when assuming the system decides to add a new user. On the other hand, if the user decides to join an existing session, the total delay is less than 800 ms, which is still acceptable (most digital TVs take longer than this when zapping through different channels).

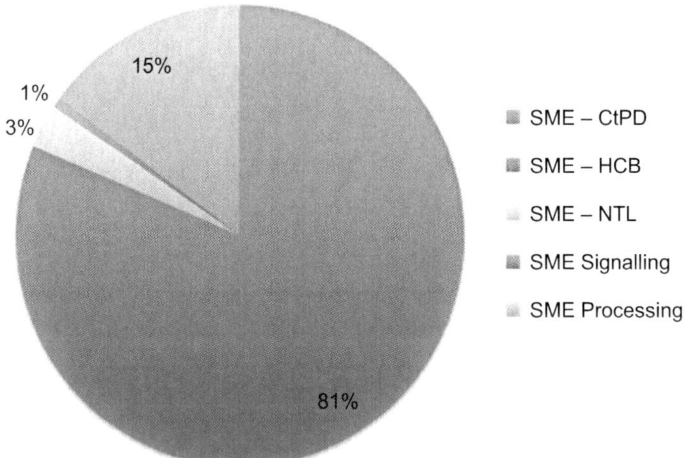

Figure 9.18 Comparison of time spent on session modification while adding a user to an existent five users single context-aware session

Another interesting fact here is that processing is significantly lower (around eight times), when compared to session initiation. This happens because the system is dealing with a single user instead of five. This means that the processing time is not uniquely related not only with the number of users (otherwise it would be around 5 times lower), but also with the scenario being considered. In addition, as previously mentioned, when considering a real-user, an extra delay of 1,400 ms would have to be added to account the real-world constraints. Please notice that this is true for the unicast scenario (the worst), as on multicast, the user terminal itself is responsible for connecting to a multicast address (provided by the SIP INVITE message).

Context change use-case

For the situation where the user context changes and according to the service rules, the multimedia delivery needs to be updated, an extra activity is involved. To get an overview of the total time between the actual context change (and not the time the SME was notified about it) and the system reaction, it is important to measure the context change itself. In this sense, around 86% of time is spent on this process (the context provider is updated, then the HCB is informed, which delivers a notification towards the SME). Figure 9.19 highlights the relationship between the amounts of time spent in the different stages of the context change scenario.

Similarly to the previous use-case, there is no need to contact the CtPD, if it is assumed that the context change does not involve any new streams. Likewise, the processing time is very less. Therefore, it is easy to assume that independently from the scenario, the system limitations are always associated with the HCB. Still, the system performs fairly well, as it is capable of reacting on context changes within 1,200 ms. When applied to a real-world situation, it would be very interesting to

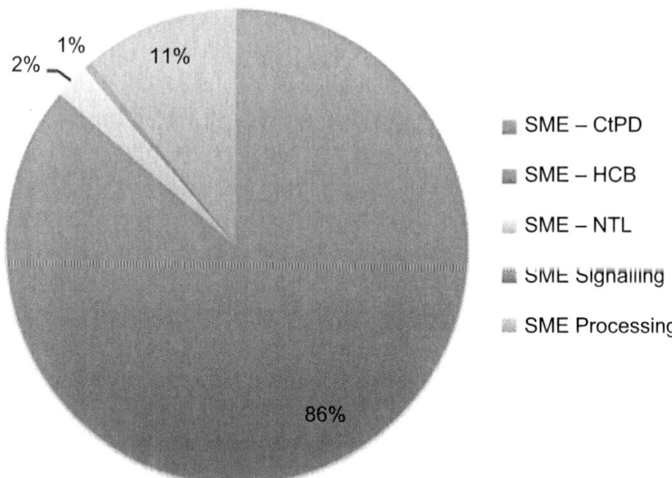

Figure 9.19 Comparison of time spent on a single user context change within an existent five users single context-aware session

understand if the users would find this amount of time as good or not (this topic is further covered in section 9.5).

9.2.3.3 Other considerations

Although not all use-cases have been covered here, the enabler specification would have allowed them. There is an exception for the mobility scenarios which have not been tested. The main reason for this relates to its dependence on the MoC component, which is out of the scope of this work. Nevertheless, if this component had been integrated in the terminal, it would have worked (in fact, this was tested within the C-CAST European Project). Nevertheless, these particular scenarios should be further explored in future work, as their impact on the quality of experience for multimedia delivery is significant (mobility).

Regarding other tests that could have been done, they would more or less fit into a combination of tests that have been performed separately throughout the entire evaluation section and therefore, are not considered in here.

Within this section, the measurements necessary to validate the non-functional requirements have been provided. By contrast, the next section focuses on the functionalities that have been implemented within the advertising system (CUCAS).

9.3 Validation of functional requirements

The functional requirements describe the capabilities that the proposed framework should implement. In this Table 9.19 section, they will be presented and classified according to three distinct but complementary categories, corresponding to the work covered in Chapters 5–7. Furthermore, an overall validation is performed to the entire CUCAS in terms of functionalities. To facilitate the validation process, the following symbols will be used to match a specific work with the respective requirement:

-	Fx2	Functionality not supported or the system does not address this issue
o	Fx3	Not implemented but the existing work could support with minor changes
+	Fx4	Partially supported but lacks full integration
++	Fx5	Addresses most or the totality of the concerns expressed

9.3.1 The human side of information technology systems

To address the challenges within this domain, we proposed the concept of a HE, an entity capable of securely managing user-related context according to user self-defined policy settings. In addition, it specified a set of adjacent modules and providers that would make this vision a reality. Before getting into the analysis, it is convenient to revise the requirements:

- *A.1*: Flexible user context and data management system.
- *A.2*: Sophisticated user context and data exposure layer.
- *A.3*: Enhanced security, privacy and trust mechanisms.

For the sake of readability and interpretation, the CUCAS is only compared with the most relevant works within each topic. In this sense, Table 9.11 describes the projects used to cross-check the requirements and validate the proposed framework.

Making the match between the list of requirements and respective works, Table 9.12 synthesizes these relationships, while the following paragraphs provide a brief explanation regarding the features supported by each work.

In what concerns the management of user-related data, almost all works introduce some sort of functionalities. However, except for the C-CAST approach (and CUCAS, which derives from it), there is no generic way to specify user-related context. Still, both SOCAM and CoBrA try to address these issues by

Table 9.11 List of most relevant works in the domain of: the human side of information technology systems

ID	Work/Project	Reference
CT	Context Toolkit	Dey et al. (2001)
CASS	Context-Aware Sub-Structure	Fahy and Clarke (2004)
SOCAM	A Middleware for Context-Aware Mobile Services	Gu et al. (2004)
CoBrA	Context Broker Architecture	Chen (2004)
C-CAST	Context to Content Casting	http://www.ict-ccast.eu/
CUCAS	Converged User-Centric Advertising System	This book

Table 9.12 Validation of most relevant works in the domain of: the human side of information technology systems

Requirement	CT	CASS	SOCAM	CoBrA	C-CAST	CUCAS
A.1	o	o	+	+	++	++
A.2	o	+	+	+	++	++
A.3	+	–	–	+	–	++
Overall	*	*	**	**	***	****

recurring to ontologies, which is not sufficient as these are limited to specific knowledge domains. In what concerns context exposure, although most works support some way of requesting data through well-defined interfaces (CT is an exception as they used agents instead of a centralized solution), except for C-CAST and CUCAS, information cannot be subscribed. Consequently, the type of applications they enable is very limited in scope, and therefore, systems are not capable of reacting on context changes. As it is possible to see, independently from the observed project, security was most of the times ignored. However, CoBrA, together with CUCAS, is an exception. They both provide users with a way to control (e.g. share and use) their privacy settings according to their situational information (e.g. where they are, who they are with and what are they doing). Nevertheless, unlike CUCAS, CoBrA did not provide any mechanisms to authenticate requests, making the system vulnerable to phishing attacks.

Notwithstanding, despite the fact that CUCAS fulfils all of the pre-defined requirements, the granularity of the privacy policies is still a concern, as each context type should be able to define their own rules, conditions and restrictions. As a workaround, for the prototype purposes, some of these were hardcoded on the PEEM component (e.g. for location it was necessary to define what means inside and outside a geographical area).

9.3.2 Understanding and predicting human behaviour

Due to the fact that the main purpose of studying this domain was adopting a methodology, in opposition to the specification of the enabler itself, the analysis and validation of the work done is bit more complex because of its abstract nature. In this sense, throughout this subsection, a distinction will be made between what was implemented and what is technically proposed by the generic reasoning enabler (RE) defined in section 6.4.2. Once again, before starting the analysis, it is important to revise the initial requirements:

- *B.1*: Generic user data modelling framework.
- *B.2*: Advanced user data reasoning framework.
- *B.3*: Classification and quantification of user-related inferred data.

Similarly to what happened in the previous subsection, some of the most relevant works in this field are described in Table 9.13. Although the main goal of this work is to generalize the understanding and prediction of human behaviour, the majority of the work available in the literature and industry focuses on specific sub-types or intermediary stages of human behaviour, such as influence, similarity or preferences. In this sense, the analysis will focus on the particularities of the RE implemented by the CUCAS.

From the Table 9.13, it is possible to see some works that were not described under the related work. The aforementioned refers to innovative commercial solutions (e.g. Klout and PeerIndex) that try to predict some metrics related to human behaviour, namely influence. In addition, and very related to the particular objectives of this work, the work done by Futureful tries to predict behaviour in a broader sense, taking into account a set of personal filters. Obviously, there are similar products on the market, but these are usually offered as a service to companies willing to pay for these predictions. In order to get a better understanding of what are the capabilities offered by each of the works introduced earlier, as well as their relation with the defined requirements, Table 9.14 presents a brief evaluation, while the following paragraphs provide an extended explanation.

Before getting into the analysis, it is important to make a disclaimer regarding the evaluation. The classification of each solution refers only to a comparison towards the defined requirements. In this sense, the same (classification) could be totally different when trying to address other issues. Furthermore, it is important to notice that most works were not developed to address an advertising specific use-case, and therefore, their main applicability refers to other domains. Another point

Table 9.13 List of most relevant works in the domain of: understanding and predicting human behaviour

ID	Work/Project	Reference
Leskovec	Patterns of influence in recommendation networks	Leskovec et al. (2006)
Bakshy	Identifying ‘influencers’ on Twitter	Bakshy et al. (2011)
Cosley	Sequential influence models in social networks	Cosley et al. (2010)
Crandall	Feedback effects between similarity and social influence in online communities	Crandall et al. (2008)
Lee	Social networks and interest similarity: The case of citeulike	Lee and Brusilovsky (2010)
Bhattacha.	Analysis of user keyword similarity in online social networks	Bhattacharyya et al. (2010)
Spertus	Evaluating similarity measures: a large-scale study in the Orkut social network	Spertus et al. (2005)
Wilkinson	Strong regularities in online peer production	Wilkinson (2008)
Goder	Consensus clustering algorithms: comparison and refinement	Goder and Filkov (2008)
Klout	The standard for influence	http://www.klout.com
PeerIndex	A measure of your online social capital	http://www.peerindex.net
Futureful	A predictive discovery engine	http://www.futureful.com
CUCAS	Converged user-centric advertising system	This book

Table 9.14 Validation of most relevant works in the domain of: understanding and predicting human behaviour

Requirement	B.1	B.2	B.3	Overall
Lescovek	+	+	o	**
Bakshy	+	o	o	**
Cosley	++	+	o	**
Crandall	++	++	+	****
Lee	+	+	–	**
Bhattacha	++	+	o	***
Spertus	++	+	o	***
Wilkinson	++	+	o	***
Goder	++	++	o	****
Klout	++	++	–	****
PeerIndex	++	++	o	****
Futureful	++	++	+	****
CUCAS	+	++	++	****

worth considering is the fact that the comparison between other works and the one we proposed will be done based on the implemented methodology and algorithms and not on the generalized reasoning framework introduced in section 6.4.2.

Just like expected, all work approaches considered some sort of data pre-processing mechanism. However, the similarities stop here. Although most considered different strategies and used different algorithms to identify the most accurate solution, the ones which have '+' on the B.1 requirement did not dig enough on this issue. One example of this situation is the CUCAS, which focused mainly on clustering, consequently not exploring other alternatives.

In what concerns the capability of combining different context information, find correlations between them and infer new data, very few were able to master this task. Nevertheless, there is an explanation for the positive sign in most of them. Despite not being able to infer or generate new knowledge about a particular user, they study user's behaviour in a broader context. This is particularly true for Lescovek, Bakshy, Cosley in what concerns influence, for Lee, Bhattacharyya and Spertus regarding similarity and for Wilkinson for the topic and popularity. From the research efforts, there are two works worth highlighting. The first is Crandall, which provides a probabilistic model based on past actions and preferences to predict the probability of future interactions. In other words, their framework allows the exploration of opinion change and behaviour adoption in social networks. Still, their work stays at a theoretical level. The second approach, which is quite relevant – Goder – introduces a methodology that allows exploring different algorithms and sources of data and arrange them according to a set of parameters. Together, these works are the inspiration for the RE belonging to the CUCAS. Taking a different approach, the commercial initiatives focus on characterizing the user specifically, rather than overviewing a community as a whole. Furthermore, by collecting data from different sources, they are able to provide more accurate and personalized results.

Regarding the last requirement, and despite some of the initiatives do provide some sort of probability or estimation for the accuracy of their results, these cannot be applied to advertising, as they cannot provide any metrics on a 'per user' basis. While this is understandable from a research point of view, this lack of information may be seen as arrogant from the commercial one. On the one hand, it is understandable that the details of their algorithmic approach are not known (i.e. secret is the soul of business). However, on the other hand, it is not comprehensible why they don't provide a metric for precision, accuracy or confidence level for their results. This is extremely important because the granularity of the information available for each individual varies and this will obviously impact the usability of this metric by any third-party provider. Besides the CUCAS, which uses the density of the cluster to assess the degree of confidence of the inference, the Futureful project uses different techniques to provide a probability to each possible event (future, as they call it). However, more details about it are not available.

Altogether, while the implementation of the RE within the CUCAS might not be the most flexible system, it is the one that provides more metrics, and the only one capable of exposing them with an accuracy value associated. Furthermore, it introduces a generic approach to study the correlation between different contexts (not only social), which was the main goal for this topic. Consequently, together with the capabilities offered by the HE and the inferences generated by the RE, the CUCAS offers advertisers the possibility to target the most influential users based on their popularity within a specific domain of influence and interests.

9.3.3 Context-aware multimedia management system

With the purpose of improving multimedia delivery for contextualized group communications across heterogeneous environments, within the CUCAS, the session management enabler (SME) was defined. Despite this component has been previously tested on the C-CAST project playground, it is important to compare it with other research initiatives, as well as existing products capable of addressing end-to-end multimedia delivery. To facilitate this analysis, below it is possible to find the previously defined requirements for this specific domain.

- *C.1*: Personalized multimedia management towards users.
- *C.2*: Efficient multimedia management towards content providers.
- ***C.3***: Flexible context-aware delivery system.
- *C.4*: Provide adaptation at session, network and transport layers.

Likewise, Table 9.15 details the most relevant works in this area. Due to its broadness, it is likely that some will focus on particular aspects of multimedia delivery, while others might cover an end-to-end solution. To better understand the main differences between the functionalities supported by each work, Table 9.16 validates at different levels, how each of the solutions fulfills the defined requirements.

At a first sight, the most relevant observation that can be made relates to the lack of research in what concerns context-awareness in this domain. More concretely, solutions that do not offer a managed environment, such as P2P Live Streaming, limit their context to the content coding, which is usually associated to

Table 9.15 List of most relevant works in the domain of: context-aware multimedia management system

ID	Work/Project	Reference
P2P Live	P2P Live Streaming websites[1]	N/A
ENAMORADO	A complete content production and delivery system in a controlled …	Papagianni et al. (2008)
Qian	Service-oriented multimedia delivery in pervasive space	Qian et al. (2009)
Lungaro	Predictive and context-aware multimedia content delivery	Lungaro et al. (2010)
Bellavista	IMS-compliant management of vertical handoffs for mobile multimedia …	Bellavista et al. (2010)
Jesus	Any-constraint personalized network selection	Jesus et al. (2008)
Chen	A new 4G architecture providing multimode terminals always best …	Chen and yang (2007)
Goergen	A session model for cross-domain interactive multi-user IPTV	Goergen et al. (2010)
CUCAS	Converged user-centric advertising system	This book

[1] http://www.justin.tv, http://www.sopcast.com, http://www.veetle.com, http://www.ustream.com

the amount of bandwidth supported by the streaming peer. Therefore, despite being popular (because it is free), this solution is the one offering less capabilities.

As expected, most works focus on the network side. Although they consider different scenarios, the most relevant research is done by Qian and Lungaro. While the former focuses on controlling each node of the network (allowing flexible network adaptation), the latter uses network context information to improve the content transmission efficiency. However, the last cannot be used for real-time scenarios. Mobility, which was a requirement, is very well addressed by the Bellavista, Chen, ENAMORADO project and Jesus. Still, Bellavista and Chen focus only on the

Table 9.16 Validation of most relevant works in the domain of: context-aware multimedia management systems

Requirement	C.1	C.2	C.3	C.4	Overall
P2P Live	o	+	–	N/A	*
ENAMORADO	+	++	o	+	***
Qian	+	+	o	++	***
Lungaro	o	o	+	++	**
Bellavista	+	+	–	+	**
Jesus	+	o	+	+	***
Chen	++	+	–	o	***
Goergen	++	++	o	o	***
CUCAS	++	++	++	+	****

handover problems. Regarding the other two, there some sort of relationship with context; however, only at session initiation. Moreover, the amount of context types considered is very less.

Regarding the control of the multimedia session, the work done by Goergen and ENAMORADO is very interesting. While most of their logic lies on the network side, they consider user context to improve the way content is prepared (or transcoded) and transmitted towards the device. Nevertheless, unfortunately, these do not account for possible adaptations that might be necessary at the network and transport layers (NTL) during an ongoing session. Making the bridge between the missing links, the CUCAS through the session management enabler, allows sessions to be updated in real-time towards the user and the content provider. In addition, it provides the necessary interfaces to trigger adaptation at theNTL, overcoming the problems previously identified. Altogether, the CUCAS is the only initiative capable of fulfilling all the requirements.

9.3.4 Advertising solution

While the previous subsections intended to validate the CUCAS outside the advertising domain, in here, the comparison will be made with products/projects that were designed to address similar requirements (i.e. designed for the future of multimedia-based advertising industry). To facilitate this analysis, here is a list of the requirements defined in section 3.3.4:

- *D.1*: Cross-converged: web, TV and mobile advertising.
- *D.2*: Multi-context advertising campaigns.
- *D.3*: User interactive and controllable advertising experience.
- *D.4*: Serendipitous advertising.
- *D.5*: Addressing users' needs, desires or intentions.
- *D.6*: Flexibility to support different business scenarios.

The advertising initiatives considered in this analysis are a sub-set of the ones presented in section 2.3 and best match the requirements. These are listed in Table 9.17. The main reason why other research initiatives are not presented in the list relates to the fact that these only address a sub-set of the problems identified and therefore do not cover an end-to-end solution. Similarly, apart from the work done by the Open Mobile Alliance (OMA) mobile advertising, the standardization contribution is inexistent in what concerns system architectures. However, the proposed architecture is used in some variations by most advertising networks, such as AdMob. Although they do not use a specific module on the devices (as specified by OMA), they provide a specific set of Application Programming Interfaces (APIs) that perform similar tasks. It is also important to notice that in this subsection, the term/project 'Hosted Solutions' refers to a particular type of advertising solutions, as described in section 2.3.3. This generalization occurs because they all offer similar capabilities. Making the match between the existing solutions and defined requirements, Table 9.18 presents the summary evaluation from a functional perspective, while the following paragraphs detail the same.

Table 9.17 List of most relevant works for an end-to-end multimedia capable advertising system

ID	Work/Project	Reference
OMA MobAd	OMA Mobile Advertising Enabler	http://www.openmobilealliance.org/technical/release_program/mobad_v1_0.aspx
Hosted Solutions	Different providers	See 2.3.3
WiPro	WIPro multi-channel ad platform http://www.wipro.com/industries/media/solutions/multichannel-ad-platform.htm	Dey and Shiroor 2010
Ericsson	Ericsson advertising broker	Asp et al., 2009
Alcatel-Lucent	Interactive IPTV advertising solution + multi-screen experience	Alcatel-Lucent, 2010; Fuerst and Fraser 2008
Comverse	Comverse mobile advertising	http://www.comverse.com/
Oracle	Oracle Communications Marketing and Advertising http://www.oracle.com/us/industries/communications/046827.html	
CUCAS	Converged User-Centric Advertising System	This book

Table 9.18 Validation of most relevant works for an end-to-end multimedia capable advertising system

Requirement	D.1	D.2	D.3	D.4	D.5	D.6	Overall
OMA MobAd	+	+	–	o	–	+	*
Hosted Solutions	o	+	–	o	–	+	**
WiPro	+	++	o	+	o	+	***
Ericsson	++	++	+	++	o	++	****
Alcatel–Lucent	++	+	o	+	–	++	***
Comverse	+	++	o	+	–	++	***
Oracle	+	+	o	+	–	++	***
CUCAS	++	++	++	++	+	++	*****

In what concerns the cross-converged systems, there is a clear tendency for all initiatives to support most mobile technologies (e.g. SMS, MMS, Wireless Application Protocol (WAP)). However, very few are capable of delivering a full featured multimedia experience (i.e. audio and video). More concretely, the only ones offering it are Ericsson, Alcatel-Lucent and CUCAS. This shows that, apart from the aforementioned exceptions, the industry is not yet capable of addressing the needs for future multimedia delivery across heterogeneous networks.

Regarding the ability to provide multi-context campaigns is where most similarities are seen. Although the granularity of the context available varies from work to work, there is a clear evidence that advertisers are willing to increase their targeting options, to methods other than demographics. This is particularly observable in the solutions offered by WiPro, Ericsson, Comverse and CUCAS, where they all possess some sort of user context or profile management component, enabling a more accurate targeting. However, none of the advertising solutions provide details on whether this advanced information (context) is uniquely used for targeting, or can also be used to trigger the advertisements (e.g. deliver ad when weather is sunny and user is near location 'x').

As explained earlier, one of the most important requirements was allowing users to control their advertising experience. Unfortunately, until the time of writing and according to the information available, only Ericsson understood this concept. Still, the options they provide are very limited, that is, they only allow users to opt-in or opt-out for the system. However, things might change in the future as most of them (e.g. WiPro, Alcatel-Lucent, Comverse and Oracle) already possess the infrastructure to support such functionalities (at least in a basic opt-in/opt-out format). On the other hand, CUCAS allows users to stay in control of their own data, assuring their privacy and security.

Slightly dependent on the previous requirement, serendipity advertising needs to take into account user requirements, otherwise it might be perceived as spam. In this sense, while most of the systems allow triggering ads without user explicit consent or demand, the existing solutions are still limited by the willingness of users to receive advertisements. This negative aspect is only tackled by Ericsson and CUCAS, which provide ways for users to select how they would consent to be targeted (e.g. commercial only on TV but discounts on mobile).

Although it is the most desirable capability an advertising system should possess, addressing user needs or intentions is still far from reality. While every solution, except for WiPro, Ericsson and CUCAS, does not cover this topic at all, the others still do it in a very conservative way. The main reason for this relates to the fact that a misjudgement might lead to a negative experience. Nevertheless, by using data mining techniques, and based on current and past context information, it is possible to predict some events (i.e. in very specific scenarios) and address them accordingly. Despite CUCAS provides mechanisms to predict the influence and similarity between users, the implementation did not account this measurement as a targeting option. However, the system is prepared to do so in the future. For the remaining (Ericsson and WiPro), although they support data mining operations, it is not perceptible how this issue could be addressed.

Finally, regarding the business models, it is clear that this is the main concern and strength of the industry. All components are designed in a generic way to allow different players to incorporate the proposed enablers in their systems. Nevertheless, the flexibility of the solutions is also dependent on the platforms and media types they support (e.g. it is not possible to have a video campaign if the platform does not support it). In fact, the main limitation of the first works (OMA MobAd, WiPro, Hosted Solutions) relates to the fact that they do not support real-time multimedia streaming (only download). Despite not being a main requirement for this work (because it focuses on the user side of advertising systems), reporting and metrics are some of the most important functionalities of an advertising system. Therefore, if it was a requirement, the CUCAS would be the only platform that would not address it.

Altogether, based on the functional requirements defined in section 3.3.4, together with Ericsson, the CUCAS is one of systems that provide the best match with the desired functionalities. Still, it has some advantages towards Ericsson in what concerns the costumer side of advertising, which is our main focus. Furthermore, it is important to stress that this validation was based on the information publicly available, plus a set of e-mails exchanged with some of these companies' marketing associates. Therefore, it is plausible that some of the assessments made for the requirements, concerning other works, might not be completely accurate. In addition, it is relevant to mention that over the 3 years during which this research was conducted, there were no significant improvements in the industry. In other words, multimedia-based advertising across heterogeneous channels is not yet a reality and the market is still supported by separated advertising silos (environments).

9.4 Validation of non-functional requirements

While the previous sections described the evaluation methodologies, the obtained results and the respective functional validation (in terms of features), in here, the components developed are validated according to the non-functional requirements previously defined in Chapter 3, Section 4. In this sense, in the following section, it is possible to find the list of requirements and respective components that fit the criteria for validation.

9.4.1 Usability

As mentioned in Chapter 3, the usability aspects are further decomposed into affordance and accessibility. Given the fact that it reflects the interaction with the user, within the defined components, this analysis is mostly suited for the advertiser interface.

9.4.1.1 Advertiser interface

As shown in Figure 8.13, this interface allows the most important interactions with the advertising system. In a single interface, by using an accordion structure, it is very simple to set up an ad with different advertising options. By using pre-defined context types, with graphical representations and functionalities' descriptions (with hover mouse), all features can be used without any specific learning or background. Likewise, by making it accessible via HTTP and a browser, it becomes very simple and easy to set up a campaign from basically anywhere. The only restriction relates to an implementation decision, which is associated with Flash/Flex. Nevertheless, most operating systems and browsers support it nowadays.

Overall: ****

9.4.1.2 Privacy context provider

Represented in Figures 8.9–8.11, this interface represents the view from the user's perspective when configuring its policy settings. In technological terms, it inherits all the advantages and disadvantages mentioned earlier (on the Advertiser Interface (AI)). However, in terms of functionality it goes one step further, as it allows users to see in real-time what kind of policies, groups and users are available for setup.

Overall: ****

9.4.1.3 Social context provider

Depicted in Figure 8.12, this interface represents one of many alike interfaces for displaying social relevant information. By using a mix of different technologies, the interactivity provided by this tool allows different commands to be executed in a single interface. Moreover, due to the possibility of applying filters to the information, the tool becomes extremely easy to use and understand.

Overall: *******

9.4.1.4 Other components

Besides the AI, it is important to notice that despite the fact that there is no apparent relation between the interfaces at the HE, RE and SME, these were thought and implemented in a way that could be used directly with the AI. In this sense, it is fair to assume that these somehow improved the usability of the aforementioned. Nevertheless, these would be easier to test if a user interface would have been provided.

Overall: ***

9.4.2 Performance

According to the definition presented in Chapter 3, performance is defined by the combination of latency, throughput, efficiency and scalability. In this sense, the following subsections provide an analysis based on the results introduced earlier in section 9.2. Due to the lack of public benchmarks (due to commercial issues), the results are validated based on common sense.

9.4.2.1 Human enabler

Although not optimal, when considering a simple request for a single user, the system performs fairly well, by being able to provide real-time context information within approximately 600 ms. Nevertheless, once the number of users, or the complexity of the requests, increases, the latency quickly arises a threshold that is not feasible for real-time services (i.e. over 15 req/s for a simple context type). Furthermore, with this increase, there is also a decrease in efficiency, as from 100 req/s, the efficiency equals zero (i.e. all requests are discarded). However, all the previous assumptions do not include any kind of security or privacy considerations.

In this sense, when authentication and authorization are introduced, there is an increase of approximately 50 ms to the overall latency. Based on the previous numbers, it is easy to conclude that this is not the limiting factor. Furthermore, this value (50 ms) remains practically constant for the rates considered in the experiments taken in section 9.2. Therefore, it is clear that under the current circumstances, the security and privacy mechanisms proposed by this work would not have problems to scale with the current configurations. Likewise, its performance is quite acceptable based on the improvements and functionalities it provides.

Overall: **

9.4.2.2 Reasoning enabler

In what concerns this component, there are different phases that need to by analysed separately. The first that concerns fetching data from the social platform is somehow independent from the developed prototype and therefore is out of scope in this analysis. Regarding the simple clustering operations, it is also important to distinguish from one-feature and multi-feature clustering. While the first shows efficiency on the matching of approximately 100% for most cases, the number of instances that are actually contained in the final clusters is much less than the initial group. On the other hand, for the multi-feature clustering, the efficiency is difficult to measure due to the limitations on the memory and processing of the machine.

In what concerns the latency, considering that these operations are related with data mining, the total time to process is rather fast (i.e. approximately 8 min for 2,000 users with 148 features). However, it is important to mention that not every type of data performs the same way (i.e. text takes longer than numbers). In addition, and probably, the most important factor is that performance is associated with the number of users, and therefore this is critical when thinking about scaling. Having that said, the solution can be considered as scalable if the machine processing and memory power can be increased without limitations. Otherwise, there is a limitation of about 10,000 users for 148 features. Still, although it allows the clustering of this amount of users, the efficiency is not the best due to the limitation resulting from the total number of cluster structures (e.g. as mentioned before, it was not possible to generate more than five clusters per consensus clustering operation).

Based on the aforementioned data, the calculations of influence and popularity are also affected. This is mostly notable in the efficiency of the reasoning process in terms of data certainty (i.e. few clusters with many users are not efficient). On the other hand, the time it takes to calculate them is less than 1 min, meaning that when the clustering structures are available, it would be rather simple to calculate one's influence or popularity within a specific group of interests/demographics (common context). In fact, when considered a single cluster structure (i.e. group of common context), the popularity calculation happens in less than 3 s in a group with 500 users. Influence on the other hand will always need to evaluate all of the generated cluster structures to assess how these evolve over time.

Overall: ***

9.4.2.3 Session management enabler

Overall, when considering the plain SME component without the context option enabled, the system is efficient and rather scalable, being capable of serving 30 parallel sessions in less than 360 ms. In addition, it performs under acceptable real-time constraints (less than 1 s) up to 250 users. In what concerns scalability, it scales more or less the same way when the number of sessions or users is increased. However, the scaling issues are different for both cases. While for the number of simultaneous sessions the complexity of processing increases (therefore causing the extra delay), the signalling is rather straightforward. On the other hand, when considering a single session (with many users), the signalling delay becomes a serious threshold to the overall latency of the session establishment.

However, when adding user context to the overall multimedia delivery process, the scalability issues quickly arise. This limitation results from the communication with the HCB only, as for a single session with five users, 60% of the setup is related with this interface. The percentage of time grows up to 80% when considering different types of session modifications (e.g. adding a new user or changing context). Therefore, while it does not scale when context is available, it still performs under acceptable conditions (this is only valid because the session is initiated by the system and therefore the delay is transparent to the users) for a small number of users.

Overall: ***

9.4.2.4 Advertiser interface

The performance associated with the AI is an uninteresting measurement to consider as it is used merely to insert advertising campaigns in the system and has no influence or impact in the system whatsoever.

9.4.3 Reliability

Reliability is usually related with three other measures: fault tolerance, robustness and security. Considering that the purpose of the implementation was a simple prototype, reliability was not a priority. Still, the overall architecture and respective implementation were thought to be capable of reducing the number of flaws, when possible. One of the main decisions towards this direction was the decentralization of the components, avoiding, when possible, single points of failure. The next subsections detail the techniques adopted.

9.4.3.1 Human enabler

Although this component is not that performant, by being the one supporting most interfaces, and consequently inputs, some extra emphasis was given to error handling. The first layer of protections is provided by a reverse proxy that acts as a firewall and at the same time takes care of load balancing (when a distributed scenario exists). Likewise, it can act as a first filter in what concerns the parameters contained within the request. Additionally, whenever a request is invalid or the context is not available (or non-existent), the component is capable of replying accordingly. In what concerns its robustness, the service has been running for more than 2 months correctly (i.e. without restarting and even after the high-load testing). The security aspects are covered by the introduction of the PEEM, Privacy Context Provider (PCP) and authentication module.

Overall: ***

9.4.3.2 Reasoning enabler

As mentioned in Chapter 6, one of the main concerns within this component is data pre-processing. Within this process, data is cleaned and prepared for mining. Although not all information is 100% accurate after this stage, it is preponderant for the overall reliability. In what concerns the methodology, by being composed by independent scripts and modules, even if one fails, the other might proceed independently. There is obviously an exception when the data input for a determined operation is dependent from a previous one. In these situations, there is nothing that can be done. As for security, it has not been considered here as this component is expected to work under a walled garden environment.

Overall: ***

9.4.3.3 Session management enabler

Just like in the previous entity, the SME has adopted a modular approach. This means that logic, service exposition, event handling and signalling are all separated. In addition to this, the component has defined interfaces to communicate with external entities to provide more granular results (e.g. NTL). In parallel, it has also incorporated default procedures, which are activated when some errors occur in a specific interface. For example, whenever context is not available, a pre-determined multimedia coding will be used. Likewise, whenever the NTL do not exist, the system is capable of performing basic sub-grouping, either by requesting network context or by using a default behaviour for each type of user. Similarly to the HE, this component has been running for 2 months without any service interruption (even after the load tests performed). This means it is capable of dealing with errors in the long term. Regarding security, as demonstrated previously, it relies on the PEEM component to assert that the advertisers managing a campaign are authorized to get access to users' context and target them accordingly.

Overall: ****

9.4.3.4 Advertiser interface

Again, this component is not worth analysing as it simply provides the interface to set up campaigns. In this sense, all the errors that might occur are usually dealt within the SME (e.g. non-existing users, etc.). Furthermore, by being a controllable interface, that is, users cannot type information themselves (when they can, the data type is checked within the graphical interface before being submitted), it helps reducing the amount of errors that need to be handled by the SME.

9.4.4 Supportability

Measures that contribute to supportability include maintainability, malleability, extensibility, portability, interoperability and testability. While the previous validations have been made on a component basis, it makes more sense in here to do it on the overall system. The main principles behind the design of the advertising framework have been its modularity and adaptability.

In other words, the system comprises three main enablers, which can act as standalone components and perform tasks independently from the others. Some examples of possible configurations have been explained in section 4.3.3. Furthermore, within each component, a modular approach has been employed, facilitating feature extension or partial modification of existing functions.

Furthermore, by using standardized web technologies (i.e. Representational State Transfer (REST) or Simple Object Access Protocol (SOAP)), the communication between different components is well described and can therefore allow particular component replacement, as long as these interfaces are respected. In addition, as most of the code is written in Java and relies on application servers, it can (theoretically) run in most operating systems. Likewise, the scripting language used for minor calculations can be easily ported to other systems.

Altogether, the system was thought as a set of enablers that when put together can support different types of advertising scenarios. Furthermore, they were also designed and implemented to work as standalone components, which means that the applicability of such entities goes well beyond the advertising use-cases as explained ahead in Chapter 10.

Overall: ****

9.5 Validation of user requirements

Although the solution validation could not be accomplished by real users, within this work, a match between the user, functional and non-functional requirements is provided. The main reason why a trial study was not possible relates to the real feeling of getting an actual discount, offer or targeted advertising experience. It is possible to emulate a promotion, but if the user cannot fulfill it, the experience cannot be fully evaluated. In this sense, the following list shows how the system addresses each of the pre-defined (in Chapter 3) user requirements. Please note that some of the functionalities might fit more than a single requirement and therefore may appear repeated.

- *Personalization* (****): this issue is addressed from different perspectives. The first relates with understanding users, by identifying which topics and people are most related with each other, simply by using their profile and social preferences. The capability of choosing who can advertise under which conditions is also a way to personalize the advertising experience.
- *Contextualization* (*****): mostly visible in multimedia delivery. In a first step, the content is adapted to the user preferences (e.g. language), terminal (e.g. supported codecs), environment (e.g. weather) and network (e.g. bandwidth). However, there is a constant monitoring and it is also capable of reacting to context changes (e.g. very noisy environment). Similar options are also available for advertisers. For a single user, it takes around 620 ms to request a single context scope and about 960 ms to detect a context change.
- *Adaptation* (****): based on the contextualization options, the system has the possibility to react to context changes in real time and adapt multimedia delivery. The adaptation occurs not only towards the client, but also on the network side. The overall time to react to context changes for single user on a local network is approximately 1,200 ms. Although it was not tested, the architecture also supports users changing terminal (device) or network while maintaining the current session.
- *Interactivity* (***): although not addressed as expected, interactivity can be achieved by using context information. Advertisers can set up an advertisement, which can be experienced in different ways according to user context. In

this sense, users could explore different context types to interact with an advertisement (e.g. when exposed to noise, the content type is changed, or when the light is reduced some special sound effect is reproduced). Again, the system would react in less than 1.5 s.

- *Privacy* (*****): by using a dedicated context provider, users can set up privacy policies that can take real-time parameters as input and can be applied to any kind of context that can be associated to a user. Furthermore, by using an authentication mechanism, it is assured that the entity requesting the information is the one authorized to do so. Altogether, the overhead introduced by such mechanisms is very low, as the overall extra time that is added to the operations is approximately 50 ms for a simple policy. However, when considering real-time context (complex policy), the same goes to 700 ms, which is still acceptable for many applications.
- *Mobility* (**): this feature was not tested because it would require theNTL, which were out of our research scope. Nevertheless, in section 7.5.5, this issue is well described and documented.
- *Relevancy* (**): by using the RE it is possible to understand, which contents are more relevant to each user and also who are the most popular peers who can influence the likelihood of some advertisement. Moreover, simply by looking what friends find interesting is by itself an interesting feature. Nevertheless, no tool was developed to associate relevance to a specific advertisement.
- *Transparency* (***): although it is not enforced, advertisers could show users which information has been used by the campaign. This is mostly beneficial to help them understand their privacy settings and also fully explore the possibilities to interact with an ad (e.g. if a user knows that its noise context is being used, it may try to experience it under different environment conditions). Furthermore, the system allows querying logs to understand which information was requested by which users (advertisers). However, there is currently no interface that shows this directly to the users.
- *Rewarding* (***): this feature could not be evaluated because the system has not been used by real users. Furthermore, the rewarding aspects are mostly associated with the offer contained within the advertisement itself, and therefore does not depend on the technology. Nevertheless, it is expected that the tools provided will enable a whole new set of rewarding experiences (e.g. need to go within 30 s to a specific location, in a very noisy environment, to redeem a discount or free bonus voucher).

Although not all the user requirements have been fulfilled with excellence, it is important to notice that according to the survey performed in the beginning of this research work, privacy, contextualization and personalization have been identified (in this order) as the three most important requirements for the future of advertising. Therefore, it is fair to acknowledge that the most important requirements have been successfully fulfilled, while the remaining have been addressed on a best effort basis. Either way, all of them have been covered by the proposed advertising framework and respective implementation.

As for future work, a simple survey should be performed to assure that the proposed functionalities are indeed addressing user needs, desires or intentions (functional requirements) within acceptable conditions (non-functional requirements). If possible, a field trial should also be realized; however, with it, there should be some sort of real advertising experience associated (discount, offer, free product or service), otherwise the obtained results would not be realistic.

9.6 Results analysis

While the previous sections provided a deep understanding of the evaluation methodologies, results and requirements validation, in here, a summary analysis is presented. To facilitate this, the analysis will focus on each of the major enabler components for the overall advertising solution.

9.6.1 Human enabler

Although this component is fundamental in terms of the functionalities that it offers, it is not without drawbacks. In the overall, its flexibility to support heterogeneous context sources is its major strength. By aggregating a policy evaluator entity (PEEM) and an authentication module, it enables privacy and security features to be added without noticeable cost (the delays introduced are very low when compared to the overall context communication processes). However, this cost aggravates when policies are complex and requires real-time context to be evaluated. This is caused by the HCB entity, which acts as a bottleneck in terms of performance and scalability.

In this sense, despite its functional design provides interesting features, its architectural design needs reengineering. However, taking into account that this component was not developed within this work, it was not possible to improve it in any way. Nevertheless, based on parallel studies, the limitations could be improved by using Not only SQL (NoSQL) databases, improved caching, indexing technologies and probably a simplification of the context schema description (currently there is a big overhead with some of the existing fields). These are all hypotheses that should be considered for future work.

9.6.2 Reasoning enabler

Although the RE introduces a methodology more than a component, there is a set of parallel initiatives that were realized through it that are worth considering. The first relates with data acquisition, which is fundamental to any data mining activity. Although the captured dataset was far from optimal, it still provided great input for the inferring process. The main reason why it cannot be publicly released relates to the initial commitment with the users to keep it private. Regarding the implementation itself (based on social network analysis), the main limitations came from memory and processing power. While in some cases (one-dimension clustering), it just limits its performance (make it slow for real-time), in others (multi-dimension clustering) it imposes a true barrier to scalability (limited to a small amount of users). In this sense, it is important to consider grid and cloud computing technologies to accommodate the aforementioned issues, where resources can scale on a 'per demand' basis.

Concerning the proposed algorithms, it has been shown that one-dimension clustering is very good for finding relationships between similar things (this is a well-known characteristic of clustering). Nevertheless, the distance algorithm and respective parameters play a preponderant role in the overall efficiency. In general, it was very accurate but still many instances were not clustered (mainly due to the dataset sparsity). Despite limited by the reasons previously mentioned, multi-dimension clustering (consensus clustering) has proved to be a very interesting technique to group user-related context, as each context possibility is treated as a binary value, and consequently the overall user context can be represented in a vectorial format. This technique should perform even better if the context types (within a scope) are well known and defined (e.g. relationship status is well represented by nine different possible values). In the future, it would be interesting to see whether these methodologies can be extended and generalized to other context types.

As for the proposed algorithms (popularity and influence), it is still early to make generalizations on its preponderance. There are many commercial initiatives trying to do this, and they are based on similar approaches. Still, they usually lack on identifying key peers under a specific set of interests or contexts (in this case). Furthermore, influence, by being derived from the first, is still dependent on its validation. Nevertheless, even when supported by an incomplete dataset, and testing only one of the different steps to identify influence, it has been shown that there is a correlation between both and it is possible to calculate. In the future, this verification should be complemented with the remaining steps and field trials (although influence will always be seen as a very uncertain metric).

9.6.3 Session management enabler

This component has shown to be rather scalable when user context is not involved. This capability comes from its modular approach, allowing the execution of several complex scenarios in a fairly simple way. However, most of them are dependent on the HCB, which has some flaws in what concerns performance and scalability. Regarding its behaviour, it was verified that when increasing the number of concurrent sessions, the processing delay increases. This happens because it acts as a Back-to-Back User Agent (B2BUA) and needs to perform several operations with other components (e.g. CtPD, NTL). On the other hand, when the number of users is increased within a session, the effects of the delay will be mostly noticeable in the signalling. This behaviour is justified by the amount of retransmissions that might occur (especially when dealing with wireless networks). As for the features introduced, the capabilities to personalize (sub-grouping), contextualize (choose content based on current context) and adapt (ability to react) multimedia delivery are among its main achievements.

Although not implemented within this work, the NTL reduce some of the complexity and processing done at the SME. In addition, they provide all of this within very reasonable time intervals. Still, it remains to be seen how would this behaviour be affected when scaling (i.e. many simultaneous users) is considered. Another limitation inherent to the load tests relates to a controlled simulated environment, where clients/terminal was emulated on a local network. In a parallel experiment (without load test), it was verified that a real device connected to a wireless network can take up to 30 times more (due to network congestion or other issues). Still, some of these problems can be overcome by connecting with the NTL to assure QoS and consequently QoE. Nevertheless, altogether, it still remains to be seen whether advertising will be moving towards real-time multimedia delivery anytime soon.

Chapter 10
Conclusions and future work

While advertising has been a solid industry for years, it is only recently that it is being explored at a real-time level. Aiming at improving the user-perceived Quality of Experience (QoE) for multimedia-based advertising across heterogeneous environments (devices, networks and contexts), this book proposes a framework based on three complementary entities, that are further explored as an architecture (see Chapter 4), composed by three main components and respective interfaces. To provide flexibility and a better targeting towards the initially identified problems, each of the components is realized as an enabler, which together form the developed *Converged User-Centric Advertising System – CUCAS*. Figure 10.1 depicts the relationship between the problems addressed and respective solutions, and gives an overall perspective.

Overall, for the scope defined for this work, CUCAS was often superior to the remaining related works (see Chapters 2 and 9). This was mostly visible when comparing supported functionalities as the non-functional settings were not possible to assess (not provided by other players). As an outcome of this work, there are the following achievements:

- The session management enabler developed was reused under several settings and use-cases within the C-CAST European project.
- Publication of more than 25 scientific papers, namely 3 book chapters, 8 journal papers and 15 conference papers (see Publications for more details).
- Spin-off to a commercial realization: Mobitto (http://www.mobitto.com).

10.1 Summary and impact

With the purpose of improving user-perceived QoE for next generation advertising, it was important to understand how was the state of the art and what were the trends for both telecommunications and Internet worlds (Simoes & Wahle, 2010). This study revealed that most works were adopting a user-centric strategy combined with some sort of multimedia technology. Using these factors as common determiner, within initial work efforts, simple architectures for personalized, group and community multimedia delivery (Al-Hezmi et al., 2003; Simoes et al., 2009b, 2009c) were prototyped, respectively.

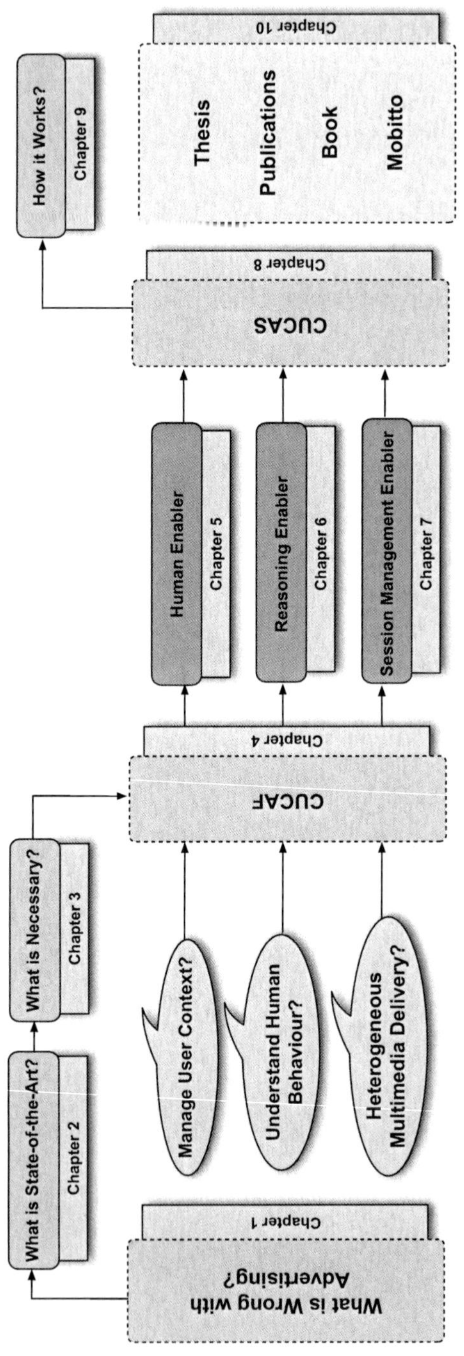

Figure 10.1 Overall perspective

Despite the usability of such approaches, at that stage, the link between future multimedia services and advertising was not established. Moreover, it was necessary to find use-cases for advertising, turning this research into something practical and applicable that could be used in a real-world scenario. Perhaps more importantly, it was crucial to understand whether the trends identified would make sense in the advertising sphere. In this sense, a survey was promoted, which was used to identify what users wanted, what they didn't like and what they would expect from future advertising initiatives (Simoes & Magedanz, 2009). It was found that users were mainly concerned with the following principles: personalization, contextualization, privacy, interactivity, mobility and adaptation. Based on the observed results, this work identified three main topics that would require further investigation, namely:

- Management of user-related data.
- Understanding of user interests, needs, wishes and concerns.
- Multimedia delivery across heterogeneous systems and devices.

Therefore, to tackle these issues, three different but complementary directions were pursued. As a consequence of such efforts, a first framework and architecture for contextualized advertising within interactive multimedia services (Simoes, 2009; Simoes et al., 2009a) was proposed, together with an example of how this system could be realized in a house of the future (Simoes & Magedanz, 2010b). In here, not only technical solutions, but also, alternative business models were proposed, namely, explaining how advertising could be used to sponsor daily user expenses, such as, water, electricity or gas.

Focusing on the first point, the main concern was finding a platform that could be capable of managing different types of user-related context (e.g. personal, social, environmental, network, etc.). To achieve such a goal, as basis of this work, the context-aware framework proposed within the C-CAST project was used. However, the former lacked security and privacy capabilities. Furthermore, the context types being used were very limited, making it difficult to understand or predict human behaviour. To overcome the aforementioned problems, both *privacy and social* (Simoes et al., 2010a) *context providers were developed*. In parallel, by extending the existing context broker, the human enabler and respective generic human profile taxonomy (Simoes et al., 2010a) are introduced. These improvements *added the missing authorization and authentication mechanisms previously non-existent on the context-aware domain*. Altogether, these changes allowed user-related data to be requested or subscribed according to a user's, or any other entity, self-defined policies. Moreover, the system-enabled dynamic policy allocation, meaning that policy parameters (e.g. location), can be based on real-time information.

Taking advantage of what was previously done, a simple approach to allow new knowledge generation and consequent exposure to third-party providers (Simoes, 2009) was proposed. Later this concept was developed further, evolving to *a methodology that allowed human behaviour understanding and prediction* (Simoes & Magedanz, 2010a, 2010c). However, despite the research showing how the methodology works, its advantages, innovations and possible business models, they do not introduce concrete (useful) results. Therefore, to test the methodology, this research work focused on concrete aspects of user behaviour. For this purpose, the user data was obtained from social networks (more specifically, Facebook). Despite the fact that the methodology was tested uniquely under social context information, it is expected that it will perform similarly for other context types (e.g. network, environmental, etc.). Then, by using the methodology introduced earlier, it was possible to:

- Discover who the most popular users are given a particular set of interests.
- Infer relationships between different user contexts.
- Understand how user preferences vary and how influence spreads according to time.

The last technical challenge of this work aimed at improving multimedia delivery within heterogeneous environments. To address this issue, the first efforts focused on *a system logic that would be capable to react on context changes*. The first prototype used a service-oriented architecture approach and was able to activate delivery through interfaces with external components (Simoes et al., 2009d). However, this first step only created *the triggering mechanism*, but had no control over the delivery itself. Besides the automatic adaptations enabled by the triggering system (e.g. noise is too high, consequently the audio will be replaced by subtitles), it was necessary to improve the network efficiency. When supported, multicast would be used, otherwise users would receive simple unicast streams. However, to support adaptation between different media types, devices, networks and user preferences, the total amount of streams that would be necessary to create to accommodate all possibilities would be huge. Therefore, this work *proposed sub-grouping* as a mechanism to address this issue (Antoniou et al., 2010a; Simoes et al., 2009a). In short, multimedia types (i.e. audio, video, subtitles) are separated into different streams (groups). Afterwards, the appropriate (according to the capabilities, network conditions and other context types) combination of the aforementioned is assigned to each user. This introduces great flexibility on the network side, but requires that the streams are synced at the client side. The overall system architecture was described in Jesus et al. (2008) and Antoniou et al. (2009), while some use-cases and capabilities were introduced in Antoniou et al. (2010c, 2010d). The validation and experimental results indicate that the system is capable of delivering personalized multimedia experiences for group communications, and, at the same time, provide the necessary personal adaptations to context changes, within acceptable time frames (Antoniou, 2010b; Coutinho et al., 2010). Altogether, *the proposed system architecture covers solutions for multi-homed systems, multiparty environments, network convergence, terminal heterogeneity and context-aware systems for real-time communications*. In other words, it showed that it is *possible to balance efficiency with personalization* (usually one compromises the other).

Altogether, this work introduces Converged User-Centric Advertising System (CUCAS), formed by a set of enablers that when combined, can be used to enrich advertising services on top of next generation networks and service delivery

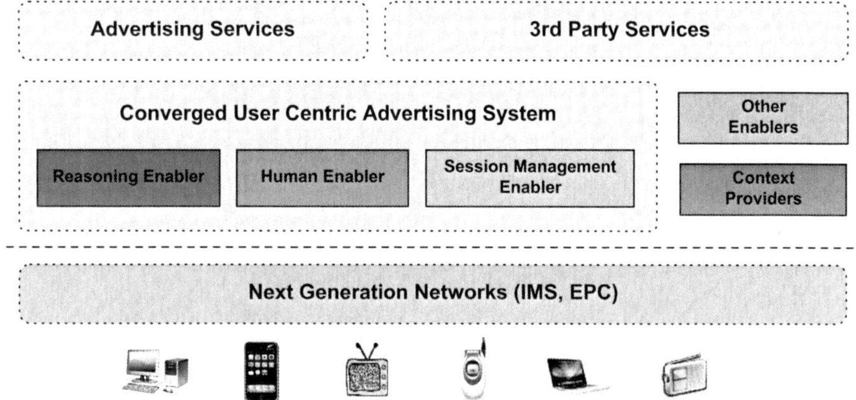

Figure 10.2 Overview of CUCAS within a service delivery platform

platforms. Furthermore, by using a set of well defined APIs, these components can even be replaced by other equivalent ones, without affecting the functions of the overall system. Figure 10.2 provides an overview of the aforementioned ecosystem.

However, contributions go beyond technical aspects. Based on the technical innovations previously mentioned, it was also possible to contribute and *improve existing business models*. Telecommunication operators, Internet or service providers can now become entities capable of aggregating and managing user context. Furthermore, different classes of service can be created according to the service being used, or the user being served. A complete list of the explored business models can be found in Moltchanov et al. (2010) and Mannweiler et al. (2010). Furthermore, it is important to mention that most of the technological developments proposed by this work are also applicable to other industries and not just advertising.

10.2 Limitations

Although this work positions itself as a pioneer for real-time user-centric multimedia-based advertising, there are three types of limitations that are still not efficiently addressed:

- *Functionalities*: although the main focus goes towards users, advertisers' needs should also be considered. In this sense, the lack of metrics and campaign statistics is a big drawback when considering a direct comparison with a commercial product. Moreover, from the set of requirements that were listed initially, there are two issues that should be better explored in the future. These are mobility (only addressed in theory) and prediction (needs further proofs).
- *Real-world deployment*: as it was possible to verify in Chapter 9, this work is far from complying with the real-world requirements in what concerns scalability and performance. Nevertheless, it is very well suited for mind shifting

and functional trials without heavy load. Still, to get a more accurate detail regarding the actual limitations and capabilities of CUCAS, the same should be tested under a cloud or grid environment without power, memory and processing restrictions.

- *Applicability*: despite the trend for mobile multimedia and advertising, the combination of both has not yet seen its true potential and may in fact never become a reality. The main reason relates with the user experience. User wants offers and discounts, not promotional videos. Besides, even if they are short and catchy enough, the time to set up such delivery is not very cost effective. In this sense, in the future, the way multimedia-based ads are consumed is yet to be discovered. Nevertheless, while some of the initial assumptions were wrong (about this tendency), the flexibility imposed at the origins of this work allows it to be explored under different domains without any type of modifications.

10.3 Outlook

Regarding future work, there are two main directions that this work could take. The first consists on focusing on the flaws encountered during our research (mentioned previously) and try to fix or improve them. The second goes one step further and should use the lessons learned from this work and apply them to an advertising solution that is applicable to a real-world scenario, that is that people are really willing to pay for it.

10.3.1 The CUCAS reengineering

Based on the aforementioned limitations, the most critical point to reengineer should be the human context broker, as by now, it is the bottleneck of the entire solution. Based on what is currently available, this change should occur not only at a technological side, but also on the concept behind it. In a first place, there should be a federated way of managing context to provide scalability and performance. Second, the overhead introduced by a verbose context description language (ContextML) should be reduced to a minimum level. Finally, the communication protocol should be tested with Extensible Messaging and Presence Protocol (XMPP), which is already used in many other applications and is known to scale and perform well when dealing with XML documents. From a technological perspective, the choice should go for non-blocking web servers, such as Node.js (http://nodejs.org) together with noSQL databases (e.g. MongoDB (http://www.mongodb.org), CouchDB (http://couchdb.apache.org) and other elastic or distributed caching mechanisms (e.g. Redis (http://redis.io), Memcached (http://memcached.org/))).

In what concerns privacy, although the enabler does allow policies to be associated with real-time context, the interface to specify this is rather complex to build. In this sense, some effort should be put into a real-time dynamic user-based policy controller. Once this is in place, the reasoning engine should be tested with other context types. While the framework has been validated for social context, it would be interesting to study how the system reacts when other sources of context are analysed together within the data mining process. In addition, the system should be flexible enough to accommodate other techniques to combine and investigate relationships (i.e. correlations) between different context types. However, to make these studies viable, this should be performed in a cloud or grid environment, where the power and processing capabilities are not limited.

Besides what has been previously mentioned, there are some things that have been identified as out of scope and should be considered for future work, namely the integration of the session management enabler with the network and transport layers. While it is not critical for the system to work, the performance and scalability of such system will need to take these issues into account. Also outside of our scope were metrics to evaluate the performance of a specific campaign. In this sense, these mechanisms should be integrated in the future to make this system more viable for commercialization. Finally, one of the points that were not deeply covered was interactivity. However, as the literature points out, this aspect is crucial to understand what users want, need or desire. Therefore, in the future, CUCAS (or any other advertising system) should implement a feedback channel to provide a two-way communication, instead of a single one.

10.3.2 The Mobitto approach

Taking into account most of the problems encountered during this work, and focusing on the applicability of an advertising solution to a real market, there are some things that need to be changed. Therefore, inspired by CUCAS, Mobitto was developed. Mobitto is a combination of both an application and a platform that aims at providing real-time targeted advertising. In short, it focuses on user QoE (just like CUCAS), and tries to change the way people consume goods, products and services.

Basically, it tries to overcome the complexity behind the CUCAS framework by offering users directly what they want. Still all the triggering and advertisements are based on user context, which is the key foundation at CUCAS. Likewise, users have the option to choose what kind of information they want to make available. However, despite it does not involve audio or video, the advertisements may contain links to content with other kinds of multimedia (which is handled directly by the terminal native operating system). For a better understanding of how the application works, Figure 10.3 presents some of the preliminary screenshots of the application.

Without going into too much detail about it, there are some points that are worth highlighting: it has a simple interface, with relevant information, but at the same time it is intuitive and offers what people want (remember: privacy and personalization). Furthermore, it was developed using the Appcelerator Titanium framework (http://www.appcelerator.com/), which allows developing a single application in Javascript and export it to native code of the leading mobile operating systems (e.g., IOS, Android, Blackberry). On the server side, some of the technologies mentioned earlier (e.g., Node.js, MongoDB, etc.) are being used to provide scalability and performance.

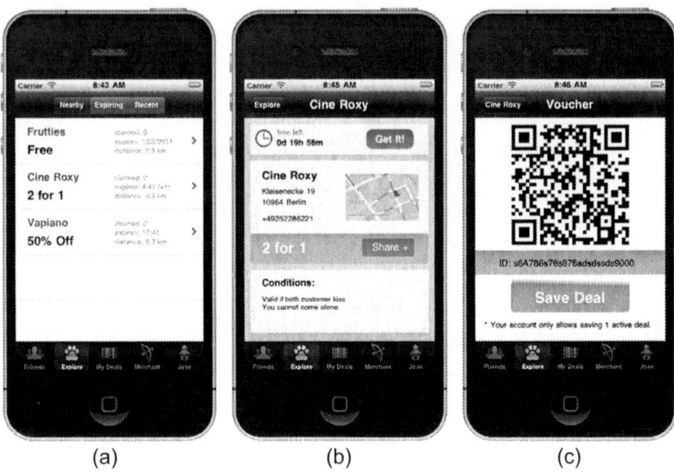

Figure 10.3 (a) Deals available, (b) Valentines offer and (c) generated voucher

10.4 Final remarks

In conclusion, while the proposed CUCAS may contribute significantly to solve the problems initially identified and address most of the requirements specified, the society is not yet willing to consume multimedia-based advertising.

Nevertheless, due to the enabler approach adopted throughout this work, the applicability of this system and respective framework goes well beyond advertising. The same has been verified within the C-CAST European project where two of the enablers (i.e. HE and SME) have been tested under different settings. Even when empowered independently, all the three enablers can extend or provide a set of features, which can be exploited in different industries. Many examples of application have been provided in the respective chapters.

Altogether, the main objectives of this work have been accomplished. While some flaws might still exist, these have been clearly identified. This not only enables future researchers to better understand existing problems, but also provides extensive information that can be useful for people and projects pretending to explore this work further (commercially or not).

Appendix

This section introduces material that supports the main book. It contains detailed information regarding specific topics.

A Advertising figures and facts

Global advertising spending in measured media is expected to exceed $500 billion for the first time ever in 2012, following an economic recovery that also sparked significant ad spending increases in 2010 (Magma Global, 2011). The same report mentions that measured Internet advertising is expected to contribute 37% of global ad growth in 2011 and is likely to reach $82 billion, a growth rate that suggests it will overtake newspaper spending (forecast at $90 billion in 2011) at some point beyond 2012. A slightly less optimistic, but still very positive report by Magma Global (2011) indicates that the overall spending will surpass the $400 billion by 2011. Figure A.1 presents both past and forecasted values.

Having that said, it is important to understand how this money will be spent by different segments. From Figure A.2, and independently from the year we analyze, it is possible to see a clear distribution of the spending across different channels.

Another interesting aspect to notice is the decrease of money being invested in print, as this is being capitalized in some sort of digital formats. Despite the general growth of the money being invested into advertising (as seen in Figure A.1), there is one sector that is particularly important to mention – mobile. With the advent of smart phones, Third generation (3G) internet, wireless hotspots and recently Fourth Generation (4G) networks, it is expected that mobile advertising steps into the market. The research conducted by Magma Global predicts a growth of 32% for 2011. Figure A.3 depicts the remaining predictions and estimated values.

But advertising is a two-sided industry. While some spend millions, others gain the same amounts from it. In fact, several companies use it as their main source of revenue. This is exactly the case for the five most visited websites in 2010 (Google.com, Facebook.com, YouTube.com, Yahoo.com, Live.com), whose main profit source is advertising. Consequently, and supported by the success seen by these companies, many others are studying the possibility to include it as a way to sponsor their services or products. This is particularly relevant because some groups of users are no longer willing to pay for services anymore.

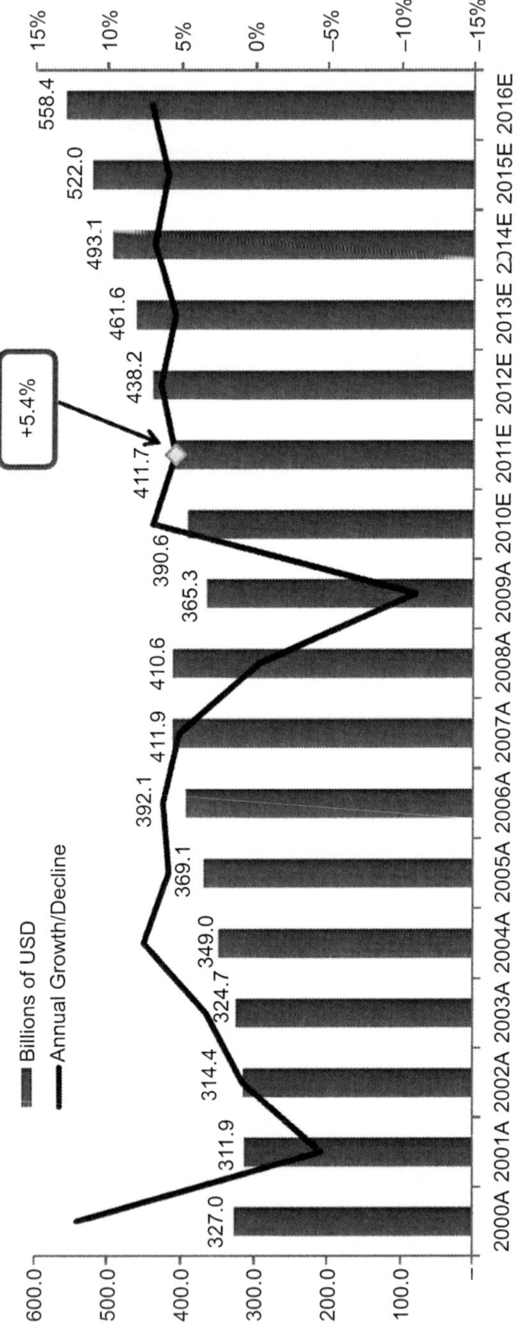

Figure A.1 Total media advertising forecast (in billions of constant USD)

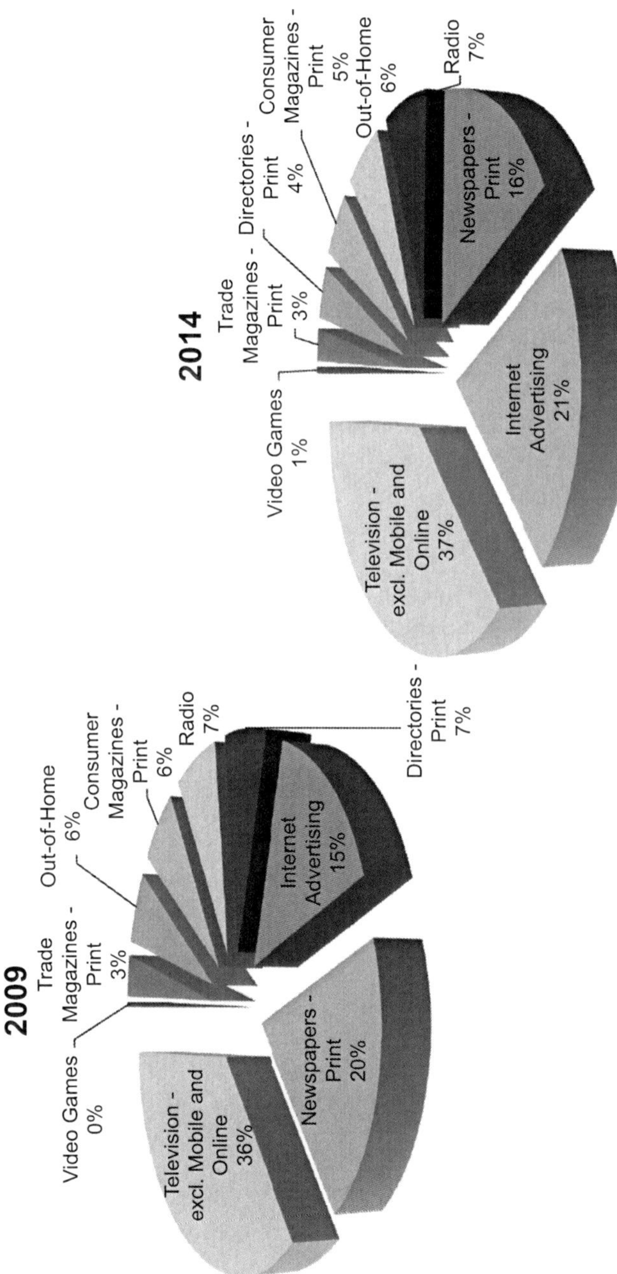

Figure A.2 Global advertising spending by segment

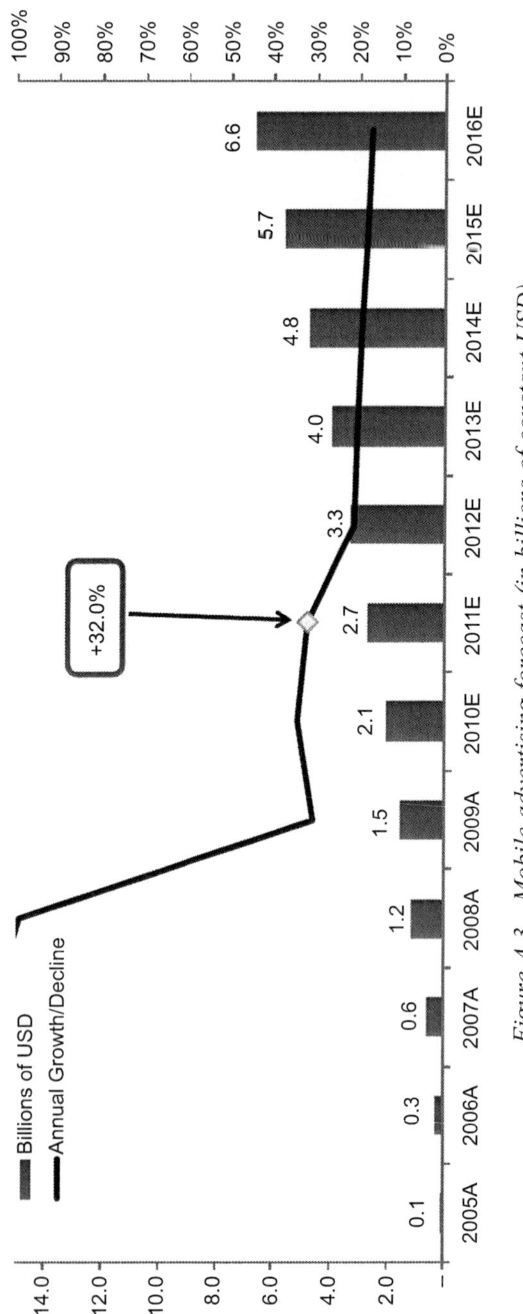

Figure A.3 Mobile advertising forecast (in billions of constant USD)

However, while advertising can be seen as necessary for economic growth, it is not without social costs. Many people criticize advertising due to hyper-commercialism, privacy invasion, violation of constitutional rights, influencing, conditioning, socio-cultural stereotyping, regulation, among other factors. For these reasons, it is important to re-think the advertising of the future, where technological advances will play a key role to overcome these and other challenges. Nevertheless, before focusing on technology, it is important to understand the consumers.

B User-generated policies

The basics of user-generated policies and their syntax have been introduced in Chapter 5. In here, some policies' examples concerning anonymization, pseudonomization and real-time user profile control are shown.

B.1 Anonymization

This situation arises when a user wants to allow third-party providers to use his/her context information, without knowing who they are. This is important to location-based services, content adaptation or many other situations we mentioned throughout the book. One easy way of doing this could be setting the 'UserProfile' context type to private. Alternatively, these settings could be applied to a single property of the 'UserProfile'. Listing B.1 exemplifies the simplest scenario.

```
<ruleset xmlns= " urn: ietf :params:xml : fokus: common-policy ">
  <rule id= r1 ">
   <conditions>
     <constraints>
       <operator match= "noregx" operandsType = "boolean " name= "equal ">
         <operand1>fn : belongsToProfile ( fn: getOrigId ( ) , 'g_jsimoes_private ' )</operand1>
         <operand2>false</ operand2>
       </operator>
     </constraints>
   </conditions>
   <actions>
   </actions>
  </rule>
</ruleset>
```

Listing B.1 Example of policy to hide user private information if request originator does not belong to 'g_jsimoes_private'

In short, if the request originator does not belong to 'g_jsimoes_private' list, the request will still be processed, but the personal information associated to the 'UserProfile' will be stripped from the response.

B.2 Pseudonomization

This case is very similar to the previous one. The only difference is that instead of removing all information regarding the user being queried, it leaves a pseudonym. This can then be used to back-trace a user on subsequent requests. However, in this case, the request originator does not know exactly who the user is. To avoid repeating the entire policy, Listing B.2 simply points out the differences (within the 'action' tag).

```
<ruleset xmlns= "urn: ietf :params:xml : fokus:
common−policy ">
    <rule id= "r1 ">
  . . .
      <actions>
        <modifyMsgParams>
          </modifyOperation name =remove
          parameter=userProfile>
        </modifyMsgParams>
      </actions>
    </rule>
</ruleset>
```

Listing B.2 Example of policy to hide user private information, except for nickname, if request originator does not belong to 'g_jsimoes_private'

B.3 User profile control

Besides the use-cases previously mentioned, there are many other things that can be achieved with the proposed solution. Although basic privacy settings are key to any system, being able to personalize them is still a challenge.

B.3.1 Example 1: Geo Location + Friend

To better understand how these rules translate into a policy, Listing B.3 represents a policy for a user A, which is composed by a ruleset of two rules. The first indicates that geoLocation can only be requested if the requestor is a friend of user A and user A is not at a pre-defined HomeLocation. The second rule applies when the requestor tries to get the list of friends of user A. In here, the only restriction is that it needs to be a friend of user A.

```
<ruleset>
  <sphere>User A</ sphere>
  <rule id= "location ">
    <conditions>
      <originatorIdentity>
        <many>
          <except_regx id= "isFriend " />
        </many>
      </originatorIdentity>
      <serviceOperation>
        <one id= "getLocation " />
      </serviceOperation>
      <validity>
      . . .
      </validity>
      <constraints>
        <operator name= "different " operandsType=
        "boolean ">
          <operand1 valueOf= "location " />
          <operand2 valueOf= "HomeLocation " />
        </operator>
      </constraints>
    </conditions>
  </rule>
  <rule id= "listFriends ">
  <conditions>
    <originatorIdentity>
      <many>
        <except_regx id= "isFriend " />
      </many>
    </originatorIdentity>
    <serviceOperation>
      <one id= "getFriendsList " />
    </serviceOperation>
    <validity>
    . . .
    </validity>
  </conditions>
  </rule>
</ruleset>
```

Listing B.3 *Example of policy that allows location to be retrieved if friend and location is different from home*

B.3.2 Example 2: Presence information

Another example that helps understanding the potential of the privacy context provider, together with the human enabler is given in Listing B.4.

This example shows how real-time context information can be retrieved and be used for the policy evaluation. In short, this policy invokes a REST service that returns the current presence status of a user. Then, this value is compared with the settings (in this case 'Available'). If it turns out that these do not match, the request is denied.

```
<ruleset>
  <rule id= "r1 ">
    <conditions>
      <originatorIdentity>
        <many />
      </originatorIdentity>
      <targetIdentity>
        <many />
      </targetIdentity>
      <validity>
        <from>2008-05-27T14:11:00 .943Z</from>
        <until>2012-05-27T14:11:00 .943Z</until>
      </validity>
      <constraints>
    <conditionalAction id="a123 type=
    "invokeRestService ">
      <attribute name= "url ">http // presence.localhost//
      getPresence/username/jsimoes</attribute>
      <attribute name= "namespace">http: //ContextML/
      1.6</attribute>
      <resultParamsRequirements>
        <operator name= "equal " operandsType= "boolean "
        match= "noregx ">
          <operand1>$a123: /ns:contextML/ ns:ctxEls/ns:
          ctxEl/ns:scope</operand1>
          <operand2>available</operand2>
        </operator>
      </resultParamsRequirements>
    </conditionalAction>
      </constraints>
    </conditions>
  </rule>
</ruleset>
```

Listing B.4 Example of policy that denies any request that is made towards a user that is not online

References

3gpp tr 23.813, 'Study on policy solutions and enhancements', v 0.1.1, January 2010. See http://www.3gpp.org/ftp/Specs/html-info/23813.htm

3gpp tr 32.808, 3rd generation partnership project: technical specification group services and system aspects; telecommunication management; study of common profile storage (cps) framework for user data for network services and management (release 8). See http://www.3gpp.org/ftp/Specs/html-info/32808.htm

3gpp ts 22.228, Signalling flows for the ip multimedia call control based on sip and sdp; stage 3. See http://www.3gpp.org/ftp/Specs/html-info/22228.htm

3gpp ts 23.002 v5.12.0, 3rd Generation partnership project; technical specification group services and systems aspects; network architecture (release 5), September 2003. See http://www.3gpp.org/ftp/Specs/html-info/23002.htm

3gpp ts 23.228 v7.6.0, 3rd Generation partnership project; technical specification group services and system aspects; ip multimedia subsystem (ims); stage 2 (release 7), December 2006. See http://www.3gpp.org/ftp/Specs/html-info/23228.htm

3gpp ts 23.401 general packet radio service (gprs) enhancements for evolved universal terrestrial radio access network (e-utran) access. See http://www.3gpp.org/ftp/Specs/html-info/23401.htm

3gpp ts 29.199-xx v8.0.0, open service access (osa); parlay x web services, September 2009. See http://www.3gpp.org/ftp/Specs/html-info/29199.htm

Acuitymobile spot relevance tm. Available from http://www.mmaglobal.com/members/acuity-mobile-inc and acquired by this company: http://www.navteq.com/

Admob. Available from http://www.admob.com/s/home/

Agarwal D., Gabrilovich E., Hall R., Josifovski V., Khanna R. 'Translating relevance scores to probabilities for contextual advertising'. 2009: 1899

Alcatel-Lucent. *Alcatel-lucent Personalized and Interactive IPTV Advertising Solution*; 2010

Al-Hezmi A., Daher G., Simoes J., Magedanz T. 'Soa-based multimedia streaming enabler over IMS networks'. *Emerging Research Projects, Applications and Services (ERPAS2009) – 7th International Conference on Advances in Mobile Computing & Multimedia*. ACM; December 2009

Antoniou J., Christophorou C., Neto A., Sargento S., Pinto F., Carapeto N., et al. 'Session and network support for autonomous context-aware multiparty communications in heterogeneous mobile systems'. *International Journal of Handheld Computing Research*. February 2010

Antoniou J., Christophorou C., Simoes J., Pitsillides A. 'Adaptive network-aided session support in context-aware converged mobile networks'. *International Journal of Autonomous and Adaptive Communication Systems*. March 2010a

Antoniou J., Christophorou C., Simoes J., Pitsillides A. 'Network-aided session management for adaptive context-aware multiparty communications'. *Book of Formal and Practical Aspects of Autonomic Computing and Networking: Specification, Development and Verification*. IGI Global; 2010b

Antoniou J., Pinto G., Simoes J., Pitsillides A. 'Supporting context-aware multiparty sessions in heterogeneous mobile networks'. *Mobile Networks and Applications*. February 2010

Antoniou J., Sargento S., Simoes J., Neto A., Christoforou C., Kellil M. 'Architecture for context-aware multiparty delivery in mobile heterogeneous networks'. *IEEE International Conference of Ultra Modern Telecommunications*. IEEE Communications Society; October 2009, p. 10

Architecture document OMA service environment, open mobile alliance (OMA), approved version 1.0.4-01, February 2007

Aslam J.A., Pavlu V., Savell R. 'A unified model for metasearch, pooling, and system evaluation'. *Proceedings of the 12th International Conference on Information and Knowledge Management*, CIKM '03. New York, NY: ACM; 2003, pp. 484–91

Asp K., Badiee F., Peng B., Svensson M., Wiss P. 'Advertising via telecom delivery channels'. *Ericsson Review*. 2009; **2**

Bagherjeiran A., Parekh R. 'Combining behavioral and social network data for online advertising'. *Proc Workshop on Data Mining in Design & Marketing in 8th Int'l Conf on Data Mining (ICDM)*. December 2008: 837–46

Baker N., Zafar M., Moltchanov B., Knappmeyer M. 'Context-aware systems and implications for future internet'. *Future Internet Conference and Technical Workshops*. May 2009

Bakshy E., Hofman J., Mason W., Watts D. *Identifying 'Influencers' on Twitter*; 2011

Baladron C., et al. 'User-centric future internet and telecommunication services'. *Towards the Future Internet*. IOS Press; 2009

Baldauf M., Dustdar S., Rosenberg F. 'A survey on context-aware systems'. *International Journal of Ad Hoc and Ubiquitous Computing*. 2007; **2**: 263–77

Bartolomeo G., Kovacikova T. 'User profile management in next generation networks'. *Proceedings of the 5th Advanced International Conference on Telecommunications*. Venice, May 2009

Barton J., Thatte S., Nielsen H. *Soap messages with attachments*. World Wide Web Consortium (W3C). December 2000. Available from http://www.w3.org/tr/soap-attachments

Bellavista P., Corradi A., Foschini L. 'Ims-compliant management of vertical handoffs for mobile multimedia session continuity'. *Communications Magazine, IEEE*. April 2010; **48**(4): 114–21

Bellavista P., Corradi A., Giannelli C. 'Mobility-aware connectivity for seamless multimedia delivery in the heterogeneous wireless internet'. *Computers and Communications, 2007. ISCC 2007. 12th IEEE Symposium on*; April 1, 2007, pp. MW–1–MW–7

Bhattacharyya P., Garg A., Wu S. 'Analysis of user keyword similarity in online social networks'. *Social Network Analysis and Mining*. 2010: 1–16. doi:10.1007/s13278-010-0006-4

Blair D. 'The history of advertising and capitalism'. *Infoart*. 1994; **6**

Blum N., Boldea I., Magedanz T., Staiger U., Stein H. 'A service broker providing real-time telecommunications services for 3rd party services'. *Computer Software and Applications Conference, Annual International*. 2009(2): 85–91

Brenner M., Wang D. 'Mining the bit pipes: Discovering and leveraging users' behavior'. *ICIN Proceedings*. October 2009: 1–6

C-CAST Project. 'D12 – specification of context casting service enablers, context management & context brokering'. *Project ICT-2007-216462*. Available from http://www.ictccast.eu/

C-CAST Project. 'D13 – specification of context detection and context-aware multi-party transport'. *Project ICT-2007-216462*. Available from http://www.ict-ccast.eu/

Chappell C. *Targeting & Personalization: The Importance of Subscriber Intelligence*. Technical report, A Lighreading Webminar, 2008

Chen H. 'An intelligent broker architecture for pervasive context-aware systems'. Phd thesis, University of Maryland, Baltimore County, December 2004

Chen Y., Yang Y. 'A new 4g architecture providing multimode terminals always best connected services'. *Wireless Communications*. 2007; **14**: 36–41

Comverse. *Get Off to a Quick Start with Comverse Mobile Advertising!* See http://www.comverse.com/data/uploads/Brochures%202010/MobileAdvertising_2010-0245.pdf

Cosley D., Huttenlocher D.P., Kleinberg J.M., Lan X., Suri S. 'Sequential influence models in social networks'. *ICWSM*. 2010

Coutand O., et al. 'Context-aware group management in mobile environments'. *IST Mobile Summit*; Germany, 2005

Coutinho N., Sargento S., Condeixa T., Janneteau C., Simoes J., Christophorou C., et al. 'Context-aware multiparty services delivery: Evaluation and experimentation'. *ICT Mobile Summit*. Florence, Italy, June 2010

Crandall D., Cosley D., Huttenlocher D., Kleinberg J., Suri, S. 'Feedback effects between similarity and social influence in online communities'. *Proceeding of the 14th ACM SIGKDD, KDD '08*. New York, NY: ACM; 2008, pp. 160–68

Damianou N. 'A policy framework for management of distributed systems'. PhD thesis, Imperial College, 2002

David L.D., Bouldin D.W. 'A cluster separation measure'. *Pattern Analysis and Machine Intelligence, IEEE Transactions on*. 1979; **PAMI-1**(2): 224–7

Dey A., Abowd G., Salber D. 'A conceptual framework and a toolkit for supporting the rapid prototyping of context-aware applications'. *Human–Computer Interaction*. December 2001; **16**: 97–166

Dey A., Shiroor R.G. 'Targeted advertising for the communication service provider'. *Intelligence in Next Generation Networks (ICIN), 2010 14th International Conference on*; 2010, pp. 1–6

Domingos P., Richardson M. 'Mining the network value of customers'. *KDD '01: Proceedings of the seventh ACM SIGKDD*. New York, NY: ACM; 2001, pp. 57–66

Drools Expert User Guide. Available from http://www.jboss.org/drools

Duarte J., Lourenco A.L., Duarte F. 'On consensus clustering validation'. *Proceedings of the 2010 joint IAPR international conference on Structural, syntactic, and statistical pattern recognition* SSPR 38, SPR'10. Berlin, Heidelberg: Springer–Verlag; 2010, pp. 385–94

Dunbar R.I.M. 'Neocortex size as a constraint on group size in primates'. *Journal of Human Evolution*. 1992; **22**(6): 469–93

Eastlake D., Hansen T. *US Secure Hash Algorithms (sha and hmac-sha)*. Internet Engineering Task Force; July 2006

Eriksson C., Akesson M. 'Ubiquitous advertising challenges'. *7th International Conference on Mobile Business*; July 2008

Eskedal T. *The Converged Telecom Market and Impact on Forecasting*. Teletronikk; 2008. ISSN 0085-7130 (document)

Ester M., Kriegel H.-P., Jörg S., Xu X. *A Density-based Algorithm for Discovering Clusters in Large Spatial Databases with Noise*; 1996

European IST-FP6 Project. *Mobilife – Wireless World Initiative*; 2006. Available from http://www.ist-mobilife.org/

European IST-FP6 Project. *Opuce – Open Platform for User-Centric Service Creation and Execution*; 2007. Available from http://www.opuce.tid.es/

European IST-FP6 Project. *Spice – Service Platform for Innovative Communication Environment*; 2008. Available from http://www.ist-spice.org/

European IST-FP7 Project. *C-cast – Context to Content Casting*; 2010. Available from http://www.ict-ccast.eu/

Fahy P., Clarke S. 'Cass – a middleware for mobile context-aware applications'. *Workshop on Context Awareness, MobiSys*; 2004

Fayyad U., Piatetsky-shapiro G., Smyth P. 'From data mining to knowledge discovery in databases'. *AI Magazine*. 1996; **17**: 37–54

Fuerst T., Fraser R. 'Enabling the multi-screen experience'. *Enriching Communications, Alcatel-Lucent*. 2008; **2** (3): 1–6

Giselle T. 'Mobile is awkward step-child of traditional channels, says analyst'. *Mobile Marketer*. 2008

Global system for mobile communications association, mobile advertising programme. Available from http://gsmworld.com/our-work/programmes-and-initiatives/mobilemedia-and-entertainment-programme/mobile-advertising/index.htm

Goder A., Filkov V. 'Consensus clustering algorithms: Comparison and refinement'. *ALENEX*. 2008: 109–17

Goergen D., Zoric J., O'Connell J., Friedrich O., Zachey B. 'A session model for cross-domain interactive multi-user iptv'. *Consumer Communications and Networking Conference (CCNC), 2010 7th IEEE*; 2010, pp. 1–6

Gomes D., Gonç andalves J.M., Santos R.O., Aguiar R. 'Xmpp based context management architecture'. *GLOBECOM Workshops (GC Wkshps)*. IEEE; 2010: 1372–77

Grady R. *Practical Software Metrics for Project Management and Process Improvement*. Prentice Hall; 1992

Greenfield J., Short K. *Software Factories: Assembling Applications with Patterns, Models, Frameworks, and Tools*. Wiley; 2004

GroupM. *This Year, Next Year*. Technical report, GroupM, 2010

Gu T., Pung H., Zhang D., Wang X. 'A middleware for building context-aware mobile services'. *Proceedings of IEEE Vehicular Technology Conference (VTC)*; 2004

Handorean R., Sen R., Hackmann G., Roman G.-C. 'Context aware session management for services in ad hoc networks'. *Services Computing, 2005 IEEE International Conference on*, vol. **1**; 2005, pp. 11–15, 113–20

Hassenzahl M., Tractinsky N. 'User experience – a research agenda'. *Behavior & Information Technology*. April 2006; **25**: 91–7

Hogg T., Lerman K. 'Social dynamics of digg'. *ICWSM*. 2010

Hourcade J., et al. 'Future internet 2020: visions of an industry expert group'. *European Commission Information Society and Media*. April 2009

Hsieh C., Liang C., Chou S. 'Personalized advertising strategy for integrated social networking websites'. *IEEE/WIC/ACM International Conference on Web Intelligence and Intelligent Agent Technology*; 2008

Hua X.-S., Mei T., Li S. 'When multimedia advertising meets the new internet era'. *IEEE 10th Int'l Workshop on Multimedia Signal Processing*. October 2008: 1–5

Hudson R. 'Cultural political economy meets global production networks: a productive meeting?' *Journal of Economic Geography*. November 2008; **8:** 421–40

IAB. *Interactive advertising bureau*. Available from http://www.iab.net/

iloop mfinity tm mobile marketing platform. Available from http://iloopmobile.com/??

ITU-T. Functional requirements and architecture of the NGN release 1. ITU-T recommendation y.2012, international telecommunications union, September 2006

Jamali M., Abolhassani H. 'Different aspects of social network analysis'. *IEEE/WIC/ACM International Conference on Web Intelligence*; December 2006, pp. 66–72

Janneteau C., Simoes J., Antoniou J., Christophorou C., Kellil M., Klein A., et al. 'Contextaware multiparty networking'. *ICT Mobile and Wireless Communications Summit*. IIMC International Information Management Corporation; June 2009. ISBN: 978-1-905824-12-0

Java community process (jcp). *jsr 289: Sip servlet v1.1. Java Specification Request (jsr) 289*. Java Community Process (jcp); August 2008. Available from http://jcp.org/en/jsr/detail?id = 289

Jesus V., Sargento S., Aguiar R. 'Any-constraint personalized network selection'. *19th International Symposium on Personal, Indoor and Mobile Radio Communications*; 2008, pp. 1–6

Josefsson S. *The base16, base32, and base64 Data Encodings, Internet Engineering task force*. October 2006

Kempe D., Kleinberg J., Tardos E. 'Maximizing the spread of influence through a social network'. *Proceedings of the 9th ACM SIGKDD*, KDD '03. New York, NY: ACM; 2003, pp. 137–46

Kempe D., Kleinberg J., Tardos E. 'Influential nodes in a diffusion model for social networks'. *ICALP*. 2005

Kern D., Harding M., Storz O., Davis N., Schmidt A. 'Shaping how advertisers see me: user views on implicit and explicit profile capture'. *CHI '08: CHI '08 extended Abstracts on Human factors in Computing Systems*. New York, NY: ACM; 2008, pp. 3363–8

Kim Y., Lee J., Park S., Choi B. *Mobile Advertisement System Using Data Push Scheduling Based on User Preference*. April 2009, pp. 1–5

Knappmeyer M., Baker N., Liaquat S., Tönjes R. 'A context provisioning framework to support pervasive and ubiquitous applications'. *Proceedings of the 4th European conference on Smart sensing and Context*, EuroSSC'09; Springer-Verlag, Berlin, Heidelberg, 2009, pp. 93–106

Knappmeyer M., Kiani S., Frà C., Moltchanov B., Baker N. 'Contextml – a light-weight context representation and context management schema' *IEEE International Symposium on Wireless Pervasive Computing*; May 2010a

Knappmeyer M., Reetz E.S., Iggena T., Brettschneider D. *Context Broker – Interface Specification*. Technical report, Mobile Communication Research, University of Applied Sciences Osnabruck, 2010b

Know Y. 'P2mp session management scheme using sip in mpls-based next generation network'. *COIN-NGNCON*. July 2006: 183–5

Kohonen T. (ed.). *Self-Organizing Maps*. Secaucus, NJ: Springer-Verlag New York, Inc.; 1997

Koponen J. 'Futureself: emerging digitized life patterns and a personal future simulation system'. *Futures, The Journal of Policy, Planning and Futures Studies*. September 2010; **42**(9): 981–94

Krumm J. 'Ubiquitous advertising: The killer application for the 21st century'. *Pervasive Computing*. 2010

Larkey L.B., Markman A.B. 'Processes of similarity judgment'. *Cognitive Science*. 2005; **29**(6):1061–76

Lee D.H., Brusilovsky P. *Social Networks and Interest Similarity: The Case of Citeulike*. ACM; 2010, pp. 151–5.

Lee K.H., Shen D., Lippman A., Reed D., Schumacher H.D. 'Connected consumption: the hidden networks of consumption'. January 2009: 1–5

Lennox J., Schulzrinne H., Rosenberg J. 'Common gateway interface for sip. rfc 3050 (proposed standard)'. January 2001

Lennox J., Schulzrinne H., Wu X. 'Call processing language (cpl): a language for user control of internet telephony services. rfc 3880 (proposed standard)'. October 2004

Leskovec J., Singh A., Kleinberg J. 'Patterns of influence in recommendation networks'. 2006: 380–9

Levine S., Kurzban R. 'Explaining clustering in social networks: towards an evolutionary theory of cascading benefits'. *Managerial and Decision Economics*. 2006; **27**: 173–87

Liapis D., Vassilaras S., Yovanof G. 'Implementing a low-cost, personalized and location based service for delivering advertisements to mobile users'. *3rd International Symposium on Wireless Pervasive Computing*; 2008

Liberal F., et al. 'Qoe and *-awareness in the future internet'. *Towards the Future Internet*. IOS Press; 2009

Lin Z., Wang L., Guo S. 'Recommendations on social network sites: from link mining perspective'. September 2009: 1–4

Lungaro P., Segall Z., Zander J. 'Predictive and context-aware multimedia content delivery for future cellular networks'. *Vehicular Technology Conference (VTC 2010-Spring), 2010 IEEE 71st*. 2010, pp. 1–5, 16–19

Luo C., Wang W., Tang J., Sun J., Li J. 'A multiparty videoconferencing system over an application-level multicast protocol'. *Multimedia, IEEE Transactions on*. December 2007; **9**(8): 1621–32

Luo M. 'Obama recasts the fund-raising landscape'. *The New York Times*, October 2008

Maes S.H. 'Pragmatic approaches to true convergence with or without ims'. *Network Operations and Management Symposium Workshops, 2008. NOMS Workshops 2008. IEEE*; 2008, pp. 44–51

Magma Global. *2011 Advertising Forecast*. Technical report, Magma Global, 2011

Mannweiler C., Moltchanov B., Simoes J. 'Context-aware smart environments enabling new business models and services'. *ITU-T Kaleidoscope – Beyond the Internet: Innovations for Future Networks and Services*; 2010

McCandless M. 'Web advertising'. *IEEE Intelligent Systems*. 1998; **13**: 8–9

McChesney R. *The Political Economy of Media: Enduring Issues, Emerging Dilemmas*. New York, NY: Monthly Review Press; May 2008, pp. 260–89. ISBN 978-158367161-0

McPherson M., Smith-Lovin L., Cook J.M. 'Birds of a feather: homophily in social networks'. *Annual Review of Sociology*. 2001; **27**(1): 415–44

Mobile entertainment forum. Available from http://www.m-e-f.org/

Mobile marketing association. Available from http://www.mmaglobal.com

Moen D. 'Overview of overlay multicast protocols'. *XOM Project*. 2004. Available from http://netlab.gmu.edu/XOM/pdfs/Multicast%20Overview.pdf

Moltchanov B., Mannweiler C., Simoes J. 'Context-awareness enabling new business models in smart spaces'. *3rd ruSMART*; St.Petersburg, Russia, August 2010

Monti S., Tamayo P., Mesirov J., Golub T. 'Consensus clustering: a resampling-based method for class discovery and visualization of gene expression microarray data'. *Machine Learning*. 2003; **52**: 91–118. doi:10.1023/A:1023949509487

Multi-channel ad platform. Available from http://www.wipro.com/industries/media/solutions/multi-channel-ad-platform.htm

Murphy P. 'The next generation networking paradigm: producer/consumer model'. *Dedicated Systems Magazine*. 2000

Nagarajan M., Baid K., Sheth A., Wang S. 'Monetizing user activity on social networks – challenges and experiences'. *Conference: Web Intelligence–WI*. September 2009; **1**: 92–9

Nicke V. '*allmählich aufwärts'. der deutsche werbemarkt 2010/2011*. Technical report, ZENTRALVERBAND DER DEUTSCHEN WERBEWIRTSCHAFT ZAW E.V, 2010

Norris V. 'The next evolution of marketing: connect with your customers by marketing with meaning'. *Journal of Advertising*. 1980; **9**

Oasis web services business process execution language. version 2.0, April 2007

Oberle K., Wahl S., Sitek A. 'Enhanced methods for sip based session mobility in a converged network'. *16th IST Mobile and Wireless Communications Summit*; 2007

Oma mobile advertising enabler. Available from http://www.openmobilealliance.org/technical/release_program/mobad_v1_0.aspx

OMA policy evaluation, enforcement and management architecture v1.0. Available from http://www.openmobilealliance.org/technical/Release_program/peem_v1_0.aspx

Open Mobile Alliance (OMA). *OMA push to talk over cellular. Candidate version 2.1, Oma*, December 2009. Available from http://www.openmobilealliance.org

Open mobile alliance, mobile advertising architecture, draft version 1.0, 23 February 2009. Available from http://www.oma.org

Oracle communications marketing and advertising. Available from http://www.oracle.com/us/industries/communications/046827.html

Organisation for economic co-operation and development (OECD). *Convergence and Next Generation Networks*. Ministerial background report, Directorate for Science, Technology and Industry (Committee for Information, Computer and Communications Policy, Korea), June 2008

Owyang J., Bernoff J., Pflaum C., Bowen E. *The Future of the Social Web*. Forrester report, April 2009

Pahlavan K., Krishnamurthy P., Hatami A., Ylianttila M., Makela J., Pichna R., et al. 'Handoff in hybrid mobile data networks'. *Personal Communication*. 2000; **7**: 34–47

Papagianni C., Tselikas N.D., Kosmatos E., Papadakis A., Keutgen T., Kaklamani D.I., et al. 'A complete content production and delivery system in a controlled multimedia network'. *Computers and Communications, 2008. ISCC 2008. IEEE Symposium on*; 2008, pp. 6–9, 44–9

Pavlovski C.J. 'Service delivery platforms in practice [ip multimedia systems (ims) infrastructure and services'. *Communications Magazine, IEEE*. March 2007; **45**(3):114–21

Performance benchmark for saml2 protocols using federation manager and access manage, Sun Microsystems, 2007

Phan T., Han J., Schneider J.-G., Ebringer T., Rogers T. 'A survey of policy-based management approaches for service oriented systems'. *ASWEC '08: Proceedings of the 19th Australian Conference on Software Engineering*. Washington, DC: IEEE Computer Society; 2008, pp. 392–401

Placecast 1020. Available from http://www.1020.com/

PricewaterhouseCoopers. *The Global Entertainment & Media Outlook: 2010–2014*. Technical report, PricewaterhouseCoopers

Qian Z., Guo M., Zhang S., Lu S. 'Service-oriented multimedia delivery in pervasive space'. *WCNC'09: Proceedings of the 2009 IEEE conference on Wireless Communications & Networking Conference*. Piscataway, NJ: IEEE Press; 2009, pp. 3012–17

Quattro wireless. Available from http://www.quattrowireless.com/

Ranganathan A., Campbell R. 'Advertising in a pervasive computing environment'. *Proceedings of the 2nd international workshop on Mobile commerce*. ACM Press; 2002

Rei K. 'A policy language for the me-centric project'. *HP Labs*. September 2002

Reich M., Liefeld T., Gould J., Lerner J., Tamayo P., Mesirov J.P. 'Genepattern 2.0'. *Nature Genetics*. 2006; **38**: 500–1

Richardson M., Domingos P. 'Mining knowledge-sharing sites for viral marketing'. *KDD '02: Proceedings of the 8th ACM SIGKDD International Conference on Knowledge Discovery and Data Mining*. New York, NY: ACM; 2002, pp. 61–70

Rosenberg J. 'A hitchhiker's guide to the session initiation protocol (sip)'. IETF, Internet Draft www.ietf.org/internet-drafts/draft-ietf-sip-hitchhikers-guide-05, Work in Progress, February 2008

Rosenberg J., Schulzrinne H., Camarillo G., Johnston A., Peterson J., Sparks R., et al. *SIP: Session initiation protocol, internet Engineering Task Force*, June 2002

Sanchez J., et al. 'Bluemail: a bluetooth-based advertisement system for commercial areas'. *3rd ACM International Workshop on Performance Monitoring, Measurement, and Evaluation of Heterogeneous Wireless and Wired Networks*; 2008

Sarwar B., Karypis G., Konstan J., Reidl J. 'Item-based collaborative filtering recommendation algorithms'. *WWW '01: Proceedings of the 10th International Conference on World Wide Web*. New York, NY: ACM; 2001, pp. 285–95

Schulzrinne & Casner. Available from http://tools.ietf.org/pdf/rfc3550.pdf; 2003

Schulzrinne et al. Available from http://tools.ietf.org/pdf/rfc3550.pdf; 1995

Schulzrinne H., et al. *Rtp: A Transport Protocol for Real-Time Applications*; www.ietf.org/rfc/rfc1889.txt, 2007

Schulzrinne H., Rao A., Lanphier R. *Real Time Streaming Protocol (RTSP), Internet Engineering Task Force*; April 1998

Schulzrinne H., Tschofenig H., Morris J., Cuellar J., Polk J. *Common policy: a document format for expressing privacy preferences*, February 2007

SDP: Session Description Protocol. Internet Engineering Task Force; June 2002

Sears J. 'The evolution of contextual advertising'. *Upgrade Magazine*. SIIA; 2005

Segaran T. *Programming Collective Intelligence: Building Smart Web 2.0 Applications*. Beijing: O'Reilly; 2007

Simoes J. 'A generic enabler framework for contextualized advertising within interactive multimedia services'. *10th IEEE International Symposium on a World of Wireless, Mobile and Multimedia Networks*; June 2009, pp 1–7. IEEE Catalog Number: CFP09WOW, Library of Congress: 2009903044, ISBN 978-1-4244-4439-7

Simoes J., Blum N., Lange L., Magedanz T. 'Mediacast for mobile communities: when the web and telecommunications converge'. *10th IEEE International Symposium on a World of Wireless, Mobile and Multimedia Networks*; June 2009b, pp. 1–7. IEEE Catalog Number: CFP09WOW, Library of Congress: 2009903044, ISBN 978-1-4244-4439-7

Simoes J., Goncalves J., Mota T., Magedanz T. 'Cats: context-aware triggering system for next generation networks'. *Wireless and Mobile Networking*, vol. 308/2009. Boston, MA: Springer; September 2009, pp. 251–62. doi:978-3-642-03840-2

Simoes J., Lamorte L., Moltchanov B., Criminisi C., Magedanz T. 'Enhanced advertising for next generation networks' in ITU-T (ed.). *ITU-T Kaleidoscope – Innovations for Digital Inclusion*. IEEE Communications Society; August 2009c

Simoes J., Lampe S., Magedanz T. 'Enabling next generation multimedia social services'. *IEEE Global Telecommunications Conference*. IEEE; 2010

Simoes J., Magedanz T. 'What happens when you mix social networks, contexts and mobile advertising in the same glass?' *International Journal of Mobile Marketing*. December 2009; **4**(2)

Simoes J., Magedanz T. 'Can you predict human behavior?'. *ICIN*; 2010a

Simoes J., Magedanz T. 'Smart advertising in the home of the future'. *International Journal of Computer Aided Engineering and Technology – Special Issue on Smart Homes Technologies and Applications*. 2010b; **2**: 164–80. doi: 10.1504/IJCAET.2010.030543

Simoes J., Magedanz T. 'Understanding and predicting human behavior for social communities'. *Handbook of Social Networks, Technologies and Applications*. Springer; 2010c

Simoes J., Magedanz T. *Contextualized User-Centric Multimedia Delivery System for Next Generation Networks*. Netherlands: Springer; 2010d, pp. 1–16

Simoes J., Riede C., Magedanz T. 'New interactive experiences with iptv services using an IMS infrastructure'. *6th Annual IEEE Consumer Communications & Networking Conference*; 2009a

Simoes J., Sargento S., Antoniou J., Christophorou C., Janneteau C., Carapeto N., et al. 'Context-aware control for personalized multiparty sessions in mobile multihomed systems'. *5th International Mobile Multimedia Communications Conference*. London: ICST/ACM; September 2009

Simoes J., Wahle S. 'The future of services in next generation networks'. *IEEE Potentials*. 2010

Simoes J., Weik P., Magedanz T. 'The human side of the future internet'. *Towards the Future Internet*. IOS Press; June 2010, pp. 183–92

Smaato open mobile advertising platform overview. Available from http://www.smaato.com/soma

Sorensen L., Skouby K. 'Next generation social networks elicitation of user requirements'. *19th International Symposium on Personal, Indoor and Mobile Radio Communications*. IEEE; September 2008, pp. 1–5

Sorge C., Girao J., Sarma A. 'Privacy-enabled identity management in the future internet'. *Towards the Future Internet*. IOS Press; 2009

Spangler W.E., Hartzel K.S., Gal-Or M. 'Exploring the privacy implications of addressable advertising and viewer profiling'. *Communication of the ACM*. 2006; **49**(5):119–23

Spertus E., Sahami M., Buyukkokten O. 'Evaluating similarity measures: a large-scale study in the Orkut social network'. *Proceedings of the eleventh ACM SIGKDD international conference on Knowledge discovery in data mining*, KDD '05. New York, NY: ACM; 2005, pp. 678–84

Strang D., Soule S.A. 'Diffusion in organizations and social movements: From hybrid corn to poison pills'. *Annual Review of Sociology*. 1998; **24**(1): 265–90

Strang T., Linnhoff-Popien C. 'A context modeling survey. Workshop on advanced context modelling, reasoning and management'. *UbiComp 2004 – The 6th International Conference on Ubiquitous Computing*. September 2004

Sun Microsystems. 'the jaintm apis: Integrated network apis for the javatm platform – a white paper describing the jain objectives, overall technical architecture and program structure'. White Paper, October 2007. Available from http://java.sun.com/products/jain

Tarnas R. *The Passion of the Western Mind: Understanding the Ideas that Have Shaped Our World View*. Ballantine Books; 1993. ISBN: 978-0345368096

Tolani R. 'The origin and evolution of advertising'. *Advertising Crossing Journal*, 2008. Available from http://www.advertisingcrossing.com Ts 25237:2008 – health informatics – pseudonymization

Tv perso free. Available from http://www.free.fr/adsl/pages/television/tv-perso.html

Uszok A., Bradshaw J., Jeffers R., Suri N., Hayes P., Breedy M. 'Kaos policy and domain services: toward a description-logic approach to policy representation, deconfliction, and enforcement'. *POLICY '03: Proceedings of the 4th IEEE International Workshop on Policies for Distributed Systems and Networks*. Washington, DC: IEEE Computer Society; 2003, p. 93

Vodafone 360. Available from http://www.theregister.co.uk/2011/10/18/vodafone_kills_360/

Weiss D., et al. 'A user profile-based personalization system for digital multimedia content'. *Proceedings of the 3rd international conference on Digital Interactive Media in Entertainment and Arts*. ACM Press; 2008

White paper on mobile commerce: opportunities and challenges. *GS1 Mobile Com – Mobile Codes Consortium*, February 2008, pp. 1–60

White paper on the future internet ppp definition. *The European Future Internet Initiative (EFII)*, January 2010, pp. 1–14

Wilkinson D.M. 'Strong regularities in online peer production'. *EC '08: Proceedings of the 9th ACM conference on Electronic commerce*. New York, NY: ACM; 2008, pp. 302–9

World wide web consortium. Available from http://www.w3c.org

Wright S., Chadha R., Lapiotis G. 'Policy-based networking'. *IEEE Network*. March 2002; **16**: 8–56

X.509: information technology – open systems interconnection – the directory: Public-key and attribute certificate frameworks, itu-t recommendation, October 2009

Yang S., Wang K. 'The influence of information sensitivity compensation on privacy concern and behavioral intention'. *SIGMIS Database*. 2009; **40**(1): 38–51

Yang W.-S., Dia J.-B., Cheng H.-C., Lin H.-T. 'Mining social networks for targeted advertising'. *Proceedings-3rd Int'l Conference on Convergence & Hybrid Information Technology, ICCIT*. January 2006; **6**: 137a–137a

Young P.H. *Individual Strategy and Social Structure: An Evolutionary Theory of Institutions*. Princeton, NJ: Princeton University Press; 1998

Yunos H.M., Gao J.Z., Shim S. 'Wireless advertising's challenges and opportunities'. *Computer*. May 2003; **36**(5): 30–7

Zadeh P.M., Moshkenani M.S. 'Mining social network for semantic advertisement'. November 2008; **1**: 611–8

Zapthink. 'The soa management landscape'. *Zapthink Bulletin*. 2006

Further reading

For more information on the topics discussed in this book, see the following articles written by the Author:

Antoniou J., Christophorou C., Neto A., Sargento S., Pinto F., Carapeto N., et al. 'Session and network support for autonomous context-aware multiparty communications in heterogeneous mobile systems'. *International Journal of Handheld Computing Research*. Feb 2010

Antoniou J., Christophorou C., Simoes J., Pitsillides A. 'Network-aided session management for adaptive context-aware multiparty communications'. *Book of Formal and Practical Aspects of Autonomic Computing and Networking: Specification, Development and Verification*. IGI Global; 2010

Antoniou J., Christophorou C., Simoes J., Pitsillides A. 'Adaptive network-aided session support in context-aware converged mobile networks'. *International Journal of Autonomous and Adaptive Communication Systems*. Mar 2010

Antoniou J., Pinto G., Simoes J., Pitsillides A. *Supporting Context-Aware Multiparty Sessions in Heterogeneous Mobile Networks, Mobile Networks and Applications*. Springer, Netherlands; Feb 2010. doi:10.1007/s11036-010-0235-9

Simoes J., Kiseleva J., Sivogolovko E., Novikov B. 'Exploring influence and interests among users within social networks'. *Social Networks: Computational Aspects and Mining*. Computer and Communication Networks Series. Springer; in press

Simoes J., Magedanz T. 'What happens when you mix social networks, contexts and mobile advertising in the same glass?' *International Journal of Mobile Marketing, Mobile Marketing Association*. Dec 2009; **4** (2)

Simoes J., Magedanz T. 'Understanding and predicting human behavior for social communities'. *Handbook of Social Networks, Technologies and Applications*. Springer; 2010

Simoes J., Magedanz T. 'Contextualized user-centric multimedia delivery system for next generation networks'. *Journal of Telecommunication Systems*. Springer, Netherlands, ISSN 1018-4864; 2010: 1–16. doi:10.1007/s11235-010-9345-8

Simoes J, Magedanz T. 'Smart advertising in the home of the future'. *International Journal of Computer Aided Engineering and Technology 2010 – Special Issue on Smart Homes Technologies and Applications*. 2010; **2**(2/3): 164–80. doi:10.1504/IJCAET.2010.030543

Simoes J., Wahle S. 'The future of services in next generation networks'. *IEEE Potentials*, Jan–Feb 2011; **30**(1). doi:10.1109MPOT.2010.938149

Simoes J., Weik P., Magedanz T. 'The human side of the future Internet'. *Towards the Future Internet*. IOS Press; 2010, pp. 183–92. doi:10.3233/978-1-60750-539-6-183

Index

Note: Page numbers followed by *f* and *t* indicate figures and tables respectively.